Early Life Origins of Human Health and Disease

Early Life Origins of Human Health and Disease

Editors

John P. Newnham Perth, W.A.

Michael G. Ross Los Angeles, Calif.

29 figures and 6 tables, 2009

KARGER Basel · Freiburg · Paris · London · New York · Bangalore · Bangkok · Shanghai · Singapore · Tokyo · Sydney

Prof. Dr. John P. Newnham
School of Women's and Infant's Health
The University of Western Australia
King Edward Memorial Hospital
Subiaco, Perth, W.A. (Australia)

Prof. Dr. Michael G. Ross
Department of Obstetrics and Gynecology
Harbor-UCLA Medical Center
Geffen School of Medicine at UCLA
and UCLA School of Public Health
Los Angeles, Calif. (USA)

Library of Congress Cataloging-in-Publication Data

Early life origins of human health and disease / editors, John P. Newnham
Michael G. Ross.
p. ; cm.
Includes bibliographical references and indexes.
ISBN 978-3-8055-9139-3 (hard cover : alk. paper)
1. Epidemiology. 2. Birth weight, Low--Health aspects. 3. Health risk assessment. 4. Prenatal influences. I. Newnham, John P. II. Ross, Michael G.
[DNLM: 1. Embryonic Development. 2. Fetal Development. 3. Disease Susceptibility. 4. Epigenesis, Genetic. 5. Prenatal Exposure Delayed Effects. 6. Risk Factors. WQ 210.5 E128 2009]
RA652.E27 2009
614.4--dc22

2009009513

Disclaimer. The statements, opinions and data contained in this publication are solely those of the individual authors and contributors and not of the publisher and the editor(s). The appearance of advertisements in the book is not a warranty, endorsement, or approval of the products or services advertised or of their effectiveness, quality or safety. The publisher and the editor(s) disclaim responsibility for any injury to persons or property resulting from any ideas, methods, instructions or products referred to in the content or advertisements.

Drug Dosage. The authors and the publisher have exerted every effort to ensure that drug selection and dosage set forth in this text are in accord with current recommendations and practice at the time of publication. However, in view of ongoing research, changes in government regulations, and the constant flow of information relating to drug therapy and drug reactions, the reader is urged to check the package insert for each drug for any change in indications and dosage and for added warnings and precautions. This is particularly important when the recommended agent is a new and/or infrequently employed drug.

www.karger.com
Printed in Switzerland on acid-free and non-aging paper (ISO 9706) by Reinhardt Druck, Basel
ISBN 978–3–8055–9139–3

Contents

Preface

The current epidemic in obesity, diabetes and the many related medical conditions presents one of our greatest ever challenges to global health. Evidence from epidemiologic, clinical and laboratory studies indicates that much of our predisposition to these chronic diseases of adulthood arises at the earliest times of life. Events before birth and environmental influences in childhood interact with our genome, modified by messages resulting from the health experiences of our ancestors. This book presents the latest evidence and concepts underpinning this exciting new field of health care. Topics include the general principles of how populations making rapid transitions to Western lifestyles are particularly at risk, how nutrition affects our development, the role of a polluted environment, implications for mental health, and the early life origins of individual diseases including obesity, diabetes and cancer. This book is essential reading for anyone interested in health care and the effects of modernisation on individuals, communities and global health.

John P. Newnham, Perth, W.A.
Michael G. Ross, Los Angeles, Calif.

Abbreviations

11βHSD2	11β hydroxysteroid dehydrpgenase type 2
ACTH	Adrenocorticotrophin
AEC	Alveolar epithelial cell
APCs	Antigen presenting cells
ARC	Arcuate nucleus
BHR	Bronchial hyperresponsiveness
BMC	Bone mineral content
BMD	Bone mineral density
BMI	Body mass index
BPA	Bisphenol A
BPD	Bronchopulmonary dysplasia
CNS	Central nervous system
CpG	regions of DNA (cytosine-phosphate-guanine)
CRH	Corticotrophin-releasing hormone
CSS	Chromosome substitution strains
DES	Diethylstilbestrol
DHAS	Dehydroepiandrosterone
DOHaD	Developmental origins of health and disease
DR	Dietary restriction
DVD	Vitamin D deficiency
DXA	Dual-energy X-ray absorptiometry
FTO	Fat mass and obesity associated gene
GC	Glucocorticoid
GH	Growth hormone
GFR	Glomerular filtration rate
GR	Glucocorticoid receptor
GWAS	Genome wide association study
HDM	House dust mite
HIF	Hypoxia-inducible factors

HPA	Hypothalamo-pituitary-adrenal axis
ICM	Inner cell mass (blastocyst)
IR	Insulin receptor
IUGR	Intrauterine growth restriction
LBW	Low birthweight
LPD	Low protein diet
MDGs	Millennium development goals
PCOS	Polycystic ovarian syndrome
PPAR	Peroxisome proliferator-activated receptor
PMNS	Pune maternal nutrition study
POMC	Proopiomelanocortin
PR	Protein restriction
RCS	Recombinant congenic strains
QTL	Quantitative trait loci
SNP	Single nucleotide polymorphism
TAG	Triacylglycerol
TGFβ	Transforming growth factor β
TLR	Toll-like receptors
VDR	Vitamin D receptor
VYSE	Visceral yolk sac endoderm

Newnham JP, Ross MG (eds): Early Life Origins of Human Health and Disease.
Basel, Karger, 2009, pp 1–10

Developmental Plasticity and the Developmental Origins of Health and Disease

Peter D. Gluckman[a] · Mark A. Hanson[b]

[a]Liggins Institute, The University of Auckland, Auckland, New Zealand; [b]Developmental Origins of Health and Disease Centre, Institute of Developmental Sciences, University of Southampton, Southampton, UK

This volume celebrates progress in research within the field of developmental origins of health and disease (DOHaD). DOHaD, as we currently recognise it, grew out of epidemiological research relating poor birth weight to later risks of metabolic and cardiovascular disease, although there is a longer history that precedes these observations. While the contributors to this volume are convinced of the importance of the field to understanding human health and disease, we must recognise that there remain significant barriers to a wider acceptance of the concept that developmental plasticity plays a major role in determining risk of later disease. There are four major issues that the field must address in order to progress.

Issue 1: The Need for a Conceptual Framework – Moving beyond Phenomenology

There is a long history of isolated indications that early life events might have long-term effects on the risks of human disease, including work in the 1970s from Dörner [1], Dörner et al. [2], Freinkel [3] and Forsdahl [4], and in the 1980s from Wadsworth et al. [5], Barker and Osmond [6] and Gennser et al. [7]. Indeed, Freinkel had argued, in his 1980 Banting lecture [3], that 'developing fetal structures may be exquisitely attuned to fine alterations in maternal fuel economy … It is suggested that concepts of teratogenesis should be expanded to include alterations occurring subsequent to organogenesis during the differentiation and proliferation of fetal cells. Such changes could cause long-range effects upon behavioural, anthropometric, and metabolic functions.'

Clues from experimental studies were also missed. Dörner, Freinkel and their groups had shown that manipulation of maternal hormonal status and nutrition in rodents had effects on the offspring. Van Assche and his group, in a series of outstanding studies [for example 8, 9], showed that experimentally induced diabetes in female rats could pass to their offspring and that growth-retarded animals had insulin resistance and abnormalities of the pancreatic islets.

But the potential significance of this work remained largely ignored until two parallel lines of clinical and epidemiological research emerged. The first concerned the effects of postnatal nutrition, and a large number of studies in the 1980s and 1990s, particularly from Lucas' group [10], related patterns of infant feeding and growth to later effects on cognitive development and metabolic function. The second arose from the now classical epidemiological studies of Barker et al. [11], who found impressive correlations between birth weight and later risk of diabetes and cardiovascular disease. Barker's group was also able to use the data from follow-up of children conceived during the Dutch Hunger Winter of 1944/45 to show that an acute nutritional insult had long-term effects which depended on the time in pregnancy when it occurred [12].

There followed a plethora of studies which reported relationships between birth weight and later blood pressure, cardiovascular disease, stroke, insulin resistance, type 2 diabetes, obesity, sarcopenia and osteoporosis, as well as schizophrenia, certain cancers, obstructive lung disease, asthma and cognitive ability. The focus remained on extremes of birth size, and one of the most striking aspects of the data, namely that the effects occurred across the entire normal range of birth weight, was largely ignored. Physiologists, including these authors, fell into a reductionist mode of trying to identify specific insults which could be linked to specific outcomes.

Eventually it was recognised that birth weight was not necessarily on the causal pathway between developmental challenge and later risk of disease. The move from 'fetal origins' to 'developmental origins' reflected this realisation, as did the addition of 'health', not just 'disease'. The challenge has been to take this raft of clinical, epidemiological and experimental data and place it within a conceptual and mechanistic framework that provides a basis for future research aimed at reducing the impact of the burden of disease.

The terms 'ultimate' and 'proximate' causes are used extensively in evolutionary biology. Proximate refers to the physiological mechanism involved in an effect, ultimate to the evolutionary pathway that allows the phenomenon to persist and become manifest [13]. The proximate mechanisms are critical to the physiologist and clinician, but a more conceptual framework encompassing ultimate causes was needed. So the question that had to be asked was 'why does early life experience become manifest in the modern world as an increased risk of metabolic and other disease?' From this follow the questions: 'why has the process evolved and why does it manifest as disease; does it contribute to the rising importance of non-communicable disease across the globe; can a unitary process explain how developmental processes contribute to metabolic disease on one hand and cognitive impairment on the other?'

Human life span is now much longer than when *Homo sapiens* evolved (mean life expectancy at birth is thought to have been in the order of 25 years in the Palaeolithic). Evolutionary processes revolve around selection for maintaining reproductive fitness, not about health or lifespan. Neel [14] was the first to suggest that the major changes in lifestyle of the modern world contributed to disease risk because they exceeded our genetically limited capacity to adapt. He argued that some populations had been selected for thrifty genes, enabling them to survive feast or famine and that such populations were now at risk in an opulent world. There are flaws in the thrifty genotype concept, but the search for thrifty genes became a dominant part of the hunt for the determinants of heart disease and diabetes. Hales and Barker [15] developed the thrifty phenotype hypothesis, which suggested that the nutritionally deprived fetus limited its growth by developing insulin resistance in order to survive and this led to a trade-off against later disease risk in an obesogenic environment. While the proximate mechanism is probably not correct, in that recent data suggest that growth-retarded infants do not develop insulin resistance until later after birth [16], this trade-off model was valuable in providing a non-genomic way in which the DOHaD concept might emerge. But it had limitations: it placed fetal growth on the causal pathway; it assumed that the change in development was always induced by signals of deprivation; it assumed a need for a severe insult or stress to the fetus.

Bateson and ourselves addressed this question by suggesting, independently [17, 18] and then collectively [19, 20], that the fetus or infant draws information from its environment and adjusts its developmental trajectory accordingly, but that in doing so there are long-term consequences. When the fetal or maternal state was particularly compromised, the maternal-fetal/infant dyad would need to make specific adjustments. These might include fetal growth retardation if nutrition was limiting with a trade-off against long-term consequences. In other words, an immediate adaptation was made and the offspring would have to 'cope with the consequences'. However, we would argue that the majority of fetuses make developmental adaptations not for immediate survival but for longer term advantage. Predictive adaptive responses are plastic 'decisions' made by the fetus/neonate in response to its environment by which it anticipates or forecasts its future from nutritional or hormonal signals from the mother (in utero or during lactation) and adjusts its phenotypic development accordingly [21]. It is important to point out that these are integrated responses affecting multiple components of the phenotype: endocrine and reproductive function, development of adipocytes and myocytes, central nervous system function, endothelial function and intermediary metabolism [22]. Plasticity has a high energetic cost and hence in general is limited to an early phase of development. However, it would appear that the plastic phase extends from conception through the weaning period in humans. Experimental evidence suggests that the windows of plasticity are not the same for different components and thus the timing of experimental insults may produce somewhat different outcomes. For example, prolonging prenatal undernutrition into the infant period in the rat can modify some aspects of the resulting metabolic phenotype [23] but not obviate the change in tempo of puberty [24].

There is not an absolute distinction between predictive and immediate adaptive responses and similar mechanisms are likely to be involved, albeit to different degrees. This model allows for both a high nutrition or low nutrition environment, and for other cues such as stress to affect later outcomes. It allows for effects within the unexceptional range of fetal exposures and it allows an explanation of why the phenomenon has been retained through evolution. Anticipatory responses are common in other taxa – for example in polyphenic species such as the African locust or the plant-hopper, the phenotype induction by early-life environmental signals occurs at a time when there can be no immediate advantage [25]. We have pointed out that developmental plasticity is a tool used across taxa to allow an individual within a species to match its phenotype to the environment it will inhabit [26]. The most important phase of life to match is that leading to and during peak reproduction. What happens later is largely reproductively and evolutionarily irrelevant. Jablonka et al. [27] have pointed out that the fidelity of the prediction need not be high for it to confer a selective advantage. While these mechanisms evolved in invertebrates and persist in vertebrates including humans, the challenge for the fetus is that its ability to read the future environment is compounded by the imperfect transduction of environmental information from mother to fetus – further, the mother may consume a diet unrepresentative of the contemporary population, or have a degree of hypertension, gestational diabetes or placental insufficiency. Thus, although mammals may have lower fidelity in their predictive adaptive responses, we would argue that the process evolved and has been sustained because it confers sufficient adaptive advantage [26].

There are generally limits on the environment the fetus can perceive, partly because there are limits on the nutrient or endocrine signals which reach it. This underlies the phenomenon of maternal constraint, which in humans effectively overrides genetic influences on fetal growth and limits the latter to match maternal size so that delivery is possible [28]. Maternal constraint may have been necessary for *Homo sapiens* to develop a large brain, allowing delivery to be timed to a point when head size was likely to be limiting, with rapid postnatal growth thereafter. But in addition it may have conferred a fitness advantage because it would always limit development to match a poorer postnatal environment than exists, giving a fail-safe phenotype. Such a phenotype is of course increasingly likely to be mismatched to the contemporary world.

While this argument has been built up around maternal undernutrition, it also applies at the other end of the spectrum. We are now facing a new phenomenon of a massive rise in the incidences of gestational diabetes and maternal obesity. From the work of Dörner and Freinkel onwards, it has been clear that these can have consequences for the offspring and they are increasingly recognised as important developmental pathways to obesity [29]. Like the mismatch pathway, in evolutionary terms they are novel phenomena.

Issue 2: The Place of Development in Biological Thought

As modern biology gained its first footings in the 18th and 19th centuries, development and embryology were considered a central component of biological research.

But by the end of the 19th century things were changing; Darwinian concepts of selection were being challenged, and the inheritance of acquired characteristics – so-called Lamarckism – was still considered a viable explanation of evolution. Early in the 20th century there was an explosion of knowledge as the particulate nature of inheritance was rediscovered, concepts of genotype and phenotype emerged, and the nucleus was identified as the organelle of inheritance. Genetic explanations became entrenched and there was little place assigned to environmental factors in determining the phenotype. By the 1920s the supremacy of genetics over developmental biology and Lamarckism was complete, and the modern synthesis between Mendelian genetics and Darwinian selection occurred in the 1930s. As evolutionary genetics become a core science, developmental environmental considerations became even more irrelevant. Selection was seen to act on the adult phenotype, derived from the inherited genotype, and how the organism had reached that point was irrelevant. The insights of Schmalhausen [30] and Waddington [31] that development played a critical part in the induction of a range of phenotypes that might emerge from a single genotype, and that selection must act on the phenotype as opposed to the genotype, were largely ignored.

When modern developmental biology re-emerged as a science, it did so largely in the narrow sense of using approaches such as knockout technologies to identify new genetic pathways for development and their function. Embryology became focused on the relatively narrow area of reproductive technologies, itself largely driven by their value to molecular biology. The advocates of the Human Genome Project claimed that they would explain the innate basis of being human and the ways in which disruptions in this process conferred risk of disease. This genomic determinism is still manifest in the enormous sums spent on seeking genetic linkages for disease aetiology [32].

In the 1960s and 1970s there was resurgence of interest in the biology of development, but from the perspective of teratology. McBride recognised that thalidomide induced limb defects and Briggs had recognised rubella embryopathy. As fetal and developmental physiology emerged, isolated observations indicated that developmental factors could influence long-term physiology without being teratogenic. Thus, the conceptual framework specific to DOHaD emerged from a growing understanding of the role of developmental processes per se.

By the end of the 20th century, evolutionary biologists recognised that selection had to act on developmental processes and the field of evolutionary developmental biology emerged [33]. Ecologists recognised that environmental factors in early life played a major part in phenotypic development. This work led to fundamentally new concepts of how evolution progresses, as West-Eberhard [34] and others argued that phenotypic processes could be assimilated and accommodated within an evolving lineage, giving new weight to concepts originally developed by Baldwin, Morgan and Osborne at the end of the 19th century. There was a reaction against genetic determinism, which on one hand accepted that DNA was but a chemical entity with the potential to code for one or many proteins (depending on splice sites and post-translational modifications), and on the other that there were direct links between genotype and phenotype. The reaction found its most extreme form in the emergence of developmental systems theory [35], a respected body of thought that argued against the supremacy of the gene and that the phenotype emerged from developmental processes operating ad seriatim on the template of inherited material.

The discovery of parental imprinting changed radically the appreciation of developmental processes. It was recognised that imprinted genes often operated on the control of growth and development, and Haig suggested from studies of the IGF-2 and IGF-2 receptor imprints in the mouse that imprinting regulated fetal growth [36]. While his concept of imprinting as a tool of maternal-paternal conflict is debatable, it nevertheless highlighted the fact that the fetus was not a passive player in its development – it might have some cards to play. The demonstration that nutrition during pregnancy could affect the expression of a mutant gene in the agouti mouse [37] led to a new focus on factors influencing development.

Jablonka and Lamb [38] had argued for some time that epigenetic processes could explain the observations that environmental influences operating in one generation can have biological echoes into subsequent generations, especially if they act during development. This may involve re-creation in each generation of the conditions which generate later effects in the life of that generation, giving the impression of genetic inheritance. For example, small mothers might generate small offspring who have small offspring through uterine size being impaired in each generation. However, there is now growing evidence for epigenetic inheritance [39, 40].

Thus, development has regained a more central place in biology. Genomics and genetics alone appear to be insufficient to explain phenotype at a number of levels, and development appeared to be more critical to evolution than had been thought from the era of the modern synthesis.

Issue 3: The Need for a Mechanism

From the first epidemiological observations, a major criticism was that of biological implausibility. How could something acting at the beginning of life have effects which were delayed until middle and old age? Without knowledge of an underlying mechanism, how could we intervene between early life and later disease risk? Once underlying mechanisms are understood, biomarkers can be developed to predict risk early in life. This would permit interventions to be better targeted but it will also allow us to get away from the use of proxy measures of development such as birthweight.

A growing amount of experimental data suggest that epigenetic processes underlie the DOHaD phenomenon. The mass of work on tissue differentiation rising from the stem cell field clearly shows that epigenetic mechanisms including DNA methylation and changes in histone structure are central to cellular differentiation. Additionally, epigenetic processes can affect the expression of genes associated with regulatory pathways through life. There are now several papers in rats [for example 41, 42] and primates [43] showing epigenetic changes in response to maternal state in genes associated with metabolism and endocrine function. In our studies in the rat we have not only shown that maternal undernutrition leads to permanent changes in expression of several genes associated with metabolic and endocrine regulation in the adult liver, but that these are underpinned by DNA methylation and histone structure changes in the promoters of these genes. In the low protein model the phenotype and expression changes are prevented by co-treatment with folate which also reverses the epigenetic change [44]. When leptin is administered to the infant rat it prevents the development of the metabolic phenotype [45] and normalises both the expression and methylation changes [46].

Issue 4: The Geopolitical Implications of DOHaD

A developmental perspective to the origins of cardiovascular and metabolic disease has widespread sociological, economic and ethical implications, but institutional barriers have hindered the integration of this work into health policy. The concept that perinatal medicine might be the appropriate place to intervene to manage adult disease, with a 30- to 50-year interval between intervention and outcome, inevitably provokes a reaction. Economic models traditionally discount benefits so far into the future. But there are signs of change. An initial economic model has been published [47] and there is now a major international project addressing this perspective [48]. With a focus on childhood obesity and impaired cognition as important outcomes of early life events, the problem of the discount rate declines given the shorter time scale over which to measure outcomes. Governments are now beginning to understand the importance of development to human health and patterns of disease. A recent UK government report [49] favoured early interventions at birth or in infancy and measurement of success over longer time frames.

Metabolic syndrome worldwide now affects 1 in 6 adults over the age of 20 years. Cardiovascular disease is still pre-eminent as a public health issue in the developed world, despite the much publicised decline in its incidence in the second half of the 20th century. It seems unlikely that this decline will be sustained, given the aging population in many developed societies. Of more concern, however, is the rising risk among young people, which may bring an upward trend in the curve over the next decade [50]. An estimated 300 million people around the world are obese (body mass index greater than 30) and 155 million school age children are overweight or obese – they may live less long than their parents. By 2020 the epidemic of chronic disease will disproportionately occur in nations such as China and India. In both populations the risk of metabolic and cardiovascular disease occurs at relatively low, by Western standards, waist-hip ratios [51, 52]. Indians at birth already have relative visceral adiposity [53]. It seems rational to postulate that the high rate of metabolic disease in these two countries has a developmental component. Birth weights are low, mothers are generally small, and the nutritional transition has been rapid. We are also seeing evidence in these populations of intergenerational switching [54]. In one generation mothers are stunted and their offspring are small and at risk of developing obesity through the mismatch pathway. In time these girls themselves may develop gestational diabetes, switching the pathway in the next generation to a hyperinsulinaemic route to obesity. In parts of Asia the incidence of gestational diabetes is now approaching or above 10% [55, 56]. We can see at one time, within one population, four intergenerational cycles – the cycle of stunted, deprived, disempowered women giving birth to stunted children; then the transition through the mismatch pathway to metabolic disease, following a degree of socioeconomic development; this is followed by two new emergent cycles of gestational diabetics giving birth to children who grow up with a higher risk of diabetes and obese mothers giving birth to children who similarly become obese. The projections for the prevalence of chronic disease in Asia are horrifying [57]. The danger is that this will have geopolitical consequences if the western lifestyle is blamed. Thus, developmental science has something to offer on a global scale.

In 2009 we still do not know the optimal way to promote a healthy life from conception to adulthood in different contexts, but a life course approach is an important and new concept. The DOHaD field is emerging from phenomenology to be a conceptually and mechanistically based science that has the tools to address the issues of finding who is most at risk, and how and when in the life cycle it is best to intervene.

References

1 Dörner G: Possible significance of pre- and/or perinatal nutrition for the pathogenesis of obesity (in German). Acta Biol Med Ger 1973;30:k19–k22.
2 Dörner G, Haller H, Leonhardt W: Possible significance of pre- and/or early postnatal nutrition in the pathogenesis of arteriosclerosis (in German). Acta Biol Med Ger 1973;31:k31–k35.
3 Freinkel N: Banting Lecture 1980: of pregnancy and progeny. Diabetes 1980;29:1023–1035.
4 Forsdahl A: Are poor living conditions in childhood and adolescence an important risk factor for arteriosclerotic heart disease? Br J Prevent Soc Med 1977;31:91–95.
5 Wadsworth M, Cripps H, Midwinter R, Colley J: Blood pressure in a national birth cohort, at the age of 36 related to social and familial factors, smoking and body mass. BMJ 1985;291:1534–1538.
6 Barker DJ, Osmond C: Infant mortality, childhood nutrition, and ischaemic heart disease in England and Wales. Lancet 1986;1:1077–1081.
7 Gennser G, Rymark P, Isberg PE: Low birth weight and risk of high blood pressure in adulthood. BMJ 1988;296:1498–1500.
8 Aerts L, Van Assche FA: Is gestational diabetes an acquired condition? J Dev Physiol 1979;1:219–225.
9 De Prins FA, Van Assche FA: Intrauterine growth retardation and development of endocrine pancreas in the experimental rat. Biol Neonat 1982;41:16–21.
10 Lucas A: Programming by early nutrition in man. Ciba Found Symp 1991;156:38–50.
11 Barker DJ, Hales CN, Fall CH, Osmond C, Phipps K, Clark PM: Type 2 (non-insulin-dependent) diabetes mellitus, hypertension and hyperlipidaemia (syndrome X): relation to reduced fetal growth. Diabetologia 1993;36:62–67.
12 Ravelli AC, van der Meulen JH, Michels RP, Osmond C, Barker DJ, Hales CN, Bleker OP: Glucose tolerance in adults after prenatal exposure to famine. Lancet 1998;351:173–177.

13 Mayr E: Cause and effect in biology. Science 1961; 134:1506.
14 Neel JV: Diabetes mellitus: a 'thrifty' genotype rendered detrimental by 'progress'? Am J Hum Genet 1962;14:353–362.
15 Hales CN, Barker DJ: Type 2 (non-insulin-dependent) diabetes mellitus: the thrifty phenotype hypothesis. Diabetologia 1992;35:595–601.
16 Mericq V, Ong KK, Bazaes RA, Pena V, Avila A, Salazar T, Soto N, Iniguez G, Dunger DB: Longitudinal changes in insulin sensitivity and secretion from birth to age three years in small- and appropriate-for-gestational-age children. Diabetologia 2005;48:2609–2614.
17 Bateson P: Fetal experience and good adult design. Int J Epidemiol 2001;30:928–934.
18 Gluckman PD, Hanson MA: Living with the past: evolution, development, and patterns of disease. Science 2004;305:1733–1736.
19 Bateson P, Barker D, Clutton-Brock T, Deb D, D'Udine B, Foley RA, Gluckman P, Godfrey K, Kirkwood T, Mirazon Lahr M, McNamara J, Metcalfe NB, Monaghan P, Spencer HG, Sultan SE: Developmental plasticity and human health. Nature 2004;430:419–421.
20 Gluckman PD, Hanson MA, Spencer HG, Bateson P: Environmental influences during development and their later consequences for health and disease: implications for the interpretation of empirical studies. Proc Royal Soc Lond B 2005;272:671–677.
21 Gluckman PD, Hanson MA, Spencer HG: Predictive adaptive responses and human evolution. Trends Ecol Evol 2005;20:527–533.
22 Gluckman PD, Hanson MA, Beedle AS: Early life events and their consequences for later disease: a life history and evolutionary perspective. Am J Hum Biol 2007;19:1–19.
23 Desai M, Gayle D, Babu J, Ross MG: Programmed obesity in intrauterine growth restricted newborns: modulation by newborn nutrition. Am J Physiol 2005;288:R91–R96.
24 Sloboda DM, Howie GJ, Vickers MH: Early-onset puberty in offspring after maternal undernutrition is exaggerated by a post-weaning high fat diet: sex specific evidence of nutritional mismatch. Early Hum Dev 2007;83(suppl 1):S65–S66.
25 Applebaum SW, Heifetz Y: Density-dependent physiological phase in insects. Annu Rev Entomol 1999; 44:317–341.
26 Gluckman PD, Hanson MA: The Fetal Matrix: Evolution, Development, and Disease. Cambridge, Cambridge University Press, 2005.
27 Jablonka E, Oborny B, Molnar I, Kisdi E, Hofbauer J, Czaran T: The adaptive advantage of phenotypic memory in changing environments. Phil Trans R Soc Lond B 1995;350:133–141.
28 Gluckman PD, Hanson MA: Maternal constraint of fetal growth and its consequences. Semin Fetal Neonatal Med 2004;9:419–425.
29 Gluckman PD, Hanson MA, Beedle AS, Raubenheimer D: Fetal and neonatal pathways to obesity. Front Horm Res 2008;36:61–72.
30 Schmalhausen IL: Factors of Evolution. New York, Blakiston (McGraw-Hill), 1949.
31 Van Speybroeck L: From epigenesis to epigenetics. The case of C.H. Waddington. Ann NY Acad Sci 2002;981:61–81.
32 The Wellcome Trust Case Control Consortium: Genome-wide association study of 14,000 cases of seven common diseases and 3,000 shared controls. Nature 2007;447:661–678.
33 Gilbert SF, Opitz JM, Raff RA: Review: resynthesizing evolutionary and developmental biology. Dev Biol 1996;173:357–372.
34 West-Eberhard MJ: Developmental Plasticity and Evolution. New York, Oxford University Press, 2003.
35 Oyama S, Griffiths PE, Gray RD: Cycles of Contingency. Developmental Systems and Evolution. Cambridge, MIT Press, 2001.
36 Haig D, Graham C: Genomic imprinting and the strange case of the insulin-like growth factor II receptor. Cell 1991;64:1045–1046.
37 Wolff GL, Kodell RL, Moore SR, Cooney CA: Maternal epigenetics and methyl supplements affect agouti gene expression in Avy/a mice. FASEB J 1998;12:949–957.
38 Jablonka E, Lamb MJ: Epigenetic Inheritance and Evolution: The Lamarckian Dimension, ed 1. Oxford, Oxford University Press, 1995.
39 Jablonka E, Lamb MJ: Evolution in Four Dimensions: Genetic, Epigenetic, Behavioral and Symbolic Variation in the History of Life. Cambridge, MIT Press/Bradford, 2005.
40 Gluckman PD, Hanson MA, Beedle AS: Non-genomic transgenerational inheritance of disease risk. Bioessays 2007;29:149–154.
41 Lillycrop KA, Slater-Jefferies JL, Hanson MA, Godfrey KM, Jackson AA, Burdge GC: Induction of altered epigenetic regulation of the hepatic glucocorticoid receptor in the offspring of rats fed a protein-restricted diet during pregnancy suggests that reduced DNA methyltransferase-1 expression is involved in impaired DNA methylation and changes in histone modifications. Br J Nutr 2007;97:1064–1073.

42 Park JH, Stoffers DA, Nicholls RD, Simmons RA: Development of type 2 diabetes following intrauterine growth retardation in rats is associated with progressive epigenetic silencing of Pdx1. J Clin Invest 2008;118:2316–2324.

43 Aagaard-Tillery KM, Grove K, Bishop J, Ke X, Fu Q, McKnight R, Lane RH: Developmental origins of disease and determinants of chromatin structure: maternal diet modifies the primate fetal epigenome. J Mol Endocrinol 2008;41:91–102.

44 Lillycrop KA, Phillips ES, Jackson AA, Hanson MA, Burdge GC: Dietary protein restriction of pregnant rats induces and folic acid supplementation prevents epigenetic modification of hepatic gene expression in the offspring. J Nutr 2005;135:1382–1386.

45 Vickers MH, Gluckman PD, Coveny AH, Hofman PL, Cutfield WS, Gertler A, Breier BH, Harris M: Neonatal leptin treatment reverses developmental programming. Endocrinology 2005;146:4211–4216.

46 Gluckman PD, Lillycrop KA, Vickers MH, Pleasants AB, Phillips ES, Beedle AS, Burdge GC, Hanson MA: Metabolic plasticity during mammalian development is directionally dependent on early nutritional status. Proc Natl Acad Sci USA 2007;104:12796–12800.

47 Alderman H, Behrman JR: Estimated economic benefits of reducing low birth weight in low-income countries. Health, nutrition and population (HNP) discussion paper of the World Bank's Human Development Network. Washington, World Bank, 2004.

48 O'Connor KC, Morton SMB, Gluckman PD: Modelling the economic antecedents and costs of low birth weight: multidimensional challenges in the international Poor Start to Life project. Early Hum Dev 2007;83(suppl 1):S39.

49 Butland B, Jebb S, Kopelman P, McPherson K, Thomas S, Mardell J, Parry V: Tackling Obesities: Future Choices (Foresight Project Report). London, Government Office for Science, 2008.

50 O'Flaherty M, Ford E, Allender S, Scarborough P, Capewell S: Coronary heart disease trends in England and Wales from 1984 to 2004: concealed levelling of mortality rates among young adults. Heart 2008;94:178–181.

51 Whincup PH, Gilg JA, Papacosta O, Seymour C, Miller GJ, Alberti KG, Cook DG: Early evidence of ethnic differences in cardiovascular risk: cross sectional comparison of British South Asian and white children. BMJ 2002;324:635.

52 Zhang XL, Shu XO, Yang G, Li HL, Cai H, Gao YT, Zheng W: Abdominal adiposity and mortality in Chinese women. Arch Int Med 2007;167:886–892.

53 Yajnik CS, Fall CHD, Coyaji KJ, Hirve SS, Rao S, Barker DJP, Joglekar C, Kellingray S: Neonatal anthropometry: the thin-fat Indian baby. The Pune maternal nutrition study. Int J Obes 2003;27:173–180.

54 Gluckman PD, Beedle AS, Hanson MA, Yap EP: Developmental perspectives on individual variation: implications for understanding nutritional needs; in Bier DM, German JB, Lonnerdal B (eds): Personalized Nutrition for the Diverse Needs of Infants and Children. Basel, Karger, 2008, pp 1–12.

55 Wong L, Tan ASA: The glucose challenge test for screening gestational diabetes in pregnant women with no risk factors. Singapore Med J 2001;42:517–521.

56 Seshiah V, Balaji V, Balaji MS, Sanjeevi CB, Green A: Gestational diabetes mellitus in India. J Assoc Physicians India 2004;52:707–711.

57 World Health Organization: Preventing Chronic Diseases: A Vital Investment. Geneva, World Health Organization, 2005.

Prof. Peter D. Gluckman
Liggins Institute, The University of Auckland
2–6 Park Avenue, Private Bag 92019
Grafton, Auckland 1023 (New Zealand)
Tel. +64 9 373 7599, ext. 86634, Fax +64 9 373 7497, E-Mail pd.gluckman@auckland.ac.nz

Newnham JP, Ross MG (eds): Early Life Origins of Human Health and Disease.
Basel, Karger, 2009, pp 11–20

Epigenetics and the Influence of Maternal Diet

Karen A. Lillycrop[a] · Mark A. Hanson[b] · Graham C. Burdge[c]

[a]Cell and Development, School of Biological Sciences, [b]DOHaD Research Division, [c]Institute of Human Nutrition, University of Southampton, Southampton, UK

Epidemiological studies show that a poor intrauterine environment induced by unbalanced maternal diet or body composition, placental insufficiency or endocrine factors induces an offspring phenotype that is characterised by an increased risk of developing chronic non-communicable diseases such as cardiovascular disease and the metabolic syndrome in later life [1]. These findings have been replicated in animal models where restricted nutrition during pregnancy induces dyslipidaemia, obesity, hypertension, hyerinsulinaemia and hyperleptinaemia in the offspring [2]. This association between poor intrauterine growth and increased risk of disease in later life may result from a predictive adaptive response where the fetus responds to environmental cues during development with permanent adjustments in its development and homeostatic systems to aid later survival and reproductive fitness. However, if these adaptations are inappropriate for the postnatal environment, they may ultimately lead to an increased risk of disease because its homeostatic capacity is mismatched to that environment [3]. The mechanism by which cues about nutrient availability in the postnatal environment are transmitted to the fetus and the process by which different, stable phenotypes are induced are beginning to be understood. The purpose of this chapter is to discuss the results of recent studies on the role of epigenetics in the induction of an altered fetal phenotype by maternal nutrition during pregnancy.

Phenotype Induction and Gene Transcription

The induction of changes to the phenotype of the offspring that persist throughout the lifespan of the organism implies stable changes to gene transcription which result in altered activities of metabolic pathways and homeostatic control processes. Feeding a protein-restricted (PR) diet during pregnancy induces reduced expression of

11β-hydroxysteroid dehydrogenase type 2 and increased expression of the glucocorticoid receptor (GR) in liver, lung, kidney and brain of the offspring during fetal, neonatal and adult life. In the liver, increased GR activity upregulates phosphoenolpyruvate carboxykinase expression and activity and so increases capacity for gluconeogenesis [2].

Restricting maternal protein intake during pregnancy and/or lactation in rats also alters the expression of genes involved in lipid homeostasis. The offspring of rats fed a PR diet during pregnancy show increased blood triacylglycerol (TAG) and non-esterified fatty acid concentrations [4]. Peroxisomal proliferator-activated receptor-α (PPARα) expression was increased in the liver of the offspring of rats fed a PR diet during pregnancy and was accompanied by upregulation of its target gene acyl-CoA oxidase, while PPARγ1 expression was unchanged. In contrast, in adipose tissue the expression of the PPARγ adipose-specific isoform $PPAR\gamma_2$ was reduced [4, 5]. Increased PPARα expression would be expected to increase TAG clearance. However, increased hepatic TAG synthesis may result from increased flux of non-esterified fatty acid from adipose tissue as a result of reduced expression of PPARγ expression [5] and may have exceeded the capacity of fatty acid clearance pathways regulated by PPARα. Overall, the offspring of dams fed a PR diet during pregnancy show impaired lipid homeostasis.

Thus, these studies demonstrate that maternal nutrient restriction during pregnancy induces long-term stable effects on transcription and, importantly, in many cases the genes which show altered expression following prenatal undernutrition are transcription factors which regulate multiple pathways in development and metabolism. Maternal nutrition thus by modifying the expression of a few key transcription factors may alter metabolic and developmental pathways leading ultimately to an altered phenotype and increased disease susceptibility.

Epigenetic Mechanisms and Regulation of Transcription

One mechanism by which maternal nutrient restriction may lead to long-term changes in gene expression within the offspring is through altered epigenetic gene regulation. As epigenetic processes are integral in determining when and where specific genes are expressed, alterations in the epigenetic regulation of genes may lead to profound phenotypic effects. The word 'epigenetics' literally means on top of genetics and refers to heritable processes which modulate gene expression potential without altering DNA sequence. The major epigenetic processes are DNA methylation, histone modification and microRNAs. As most is understood about DNA methylation, this will be the main focus of this chapter.

Methylation at the 5′ position of cytosine in DNA within a CpG dinucleotide (the p denotes the intervening phosphate group) is a common modification in mammalian genomes and constitutes a stable epigenetic mark that is transmitted through cell division [6]. CpG dinucleotides are found clustered at the 5′ ends of genes/promoters in

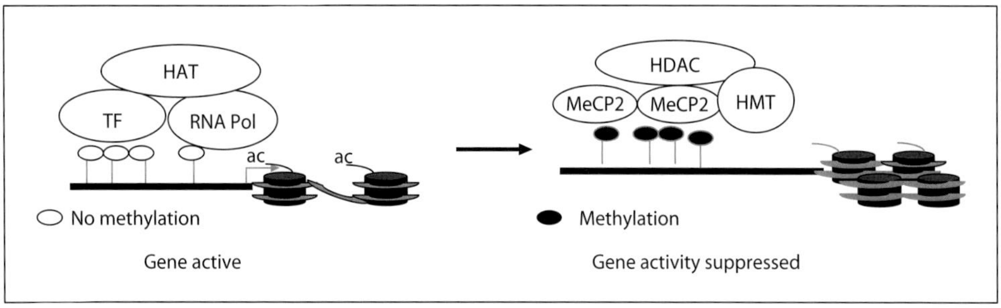

Fig. 1. Regulation of transcription by DNA methylation. When CpG dinucleotides are unmethylated, transcription factors bind, recruiting histone acetyl transferases (HATs); this leads to RNA polymerase binding and transcription of the gene. Methylation of CpGs by the activity of Dnmt blocks transcription factor binding and leads to the recruitment of MeCP2, which in turn recruits the HDAC/histone methyl transferase (HMT) complex. The overall effect of DNA and histone methylation is to induce long-term silencing of transcription.

regions known as CpG islands. Hypermethylation of these CpG islands is associated with transcriptional repression, while hypomethylation of CpG islands is associated with transcriptional activation. DNA methylation can induce transcriptional silencing by blocking the binding of transcription factors and/or through promoting the binding of the methyl CpG-binding protein (MeCP2). The latter binds to methylated cytosines and, in turn, recruits histone-modifying complexes to the DNA [7]. MeCP2 recruits both histone deacetylases (HDACs), which remove acetyl groups from the histones, and histone methyl transferases which methylate lysine 9 on H3, resulting in a closed chromatin structure and transcriptional silencing [7]. Covalent modifications to histones, such as acetylation and methylation, alter chromatin structure [8] and hence the ability of the transcriptional machinery to gain access to the DNA (fig. 1).

DNA methylation is important for asymmetrical silencing of imprinted genes, X chromosome inactivation and silencing of retrotransposons [6]. DNA methylation also plays a key role in cell differentiation by silencing the expression of specific genes during the development and differentiation of individual tissues. The methylation of CpGs is largely established during embryogenesis or in early postnatal life. Following fertilisation, maternal and paternal genomes undergo extensive demethylation. Demethylation is an active process that strips the male genome of methylation within hours of fertilisation; by contrast the maternal genome is only passively demethylated during subsequent cleavage divisions. Thus embryonic DNA becomes hypomethylated, which correlates with the pluripotency of these embryonic cells. Imprinted genes escape this erasure. This period of demethylation is followed by global de novo methylation just prior to blastocyst implantation during which 70% of CpGs are methylated, mainly in repressed heterochromatin regions and in repetitive sequences such as retrotransposable elements [6]. In addition, during development and early postnatal life de novo methylation also occurs of tissue specific genes, limiting gene expression and the developmental fates of differentiating cells. For example, the expression of Oct-4, a key regulator of cellular pluripotency in the early embryo, is permanently silenced by hypermethylation around E6.5 in the mouse [9], while HoxA5 and HoxB5 which are required for later stages of development are methylated and silenced in early postnatal life [10]. However, once these methylation patterns have been established during development, these epigenetic markers are in most cases maintained with high fidelity throughout life. The periods during development when these methylation patterns are being established are likely to be susceptible to early life environmental influences.

Early Life Environment and Epigenetic Gene Regulation

There is increasing evidence that prenatal and early postnatal environments can modify the epigenetic regulation of specific genes. Pham et al. [11] have shown that ligation of a uterine artery in the rat leads to increased p53 methylation and decreased p53 expression in the kidney of the offspring, which is associated with increased apoptosis and reduced nephron number. Variations in maternal behaviour have also been shown to lead to epigenetic changes in rats. Weaver et al. [12] showed that pups raised by rat dams which showed poorer nurturing had an increased stress response. The effect was due to hypermethylation of a specific CpG within the promoter of the GR gene in the hippocampus of the offspring. These changes persisted into adulthood and were associated with altered histone acetylation and reduced binding of the transcription factor NGF1A to the GR promoter. Central infusion of the HDAC inhibitor trichostatin A removed the differences in histone acetylation, DNA methylation, NGF1A binding and the hypothalamus-pituitary-adrenal axis stress response.

The effects of early nutrition on the epigenetic regulation of imprinted genes and intracisternal A particle (IAP) retrotransposons have also been reported. Mouse embryos cultured in Whitten's medium without amino acids showed bi-allelic expression of the imprinted H19 gene, while those cultured in medium containing amino acids showed mono-allelic expression [13]. In humans, assisted reproductive technologies are associated with increased risk of Angelman's syndrome and Beckwith-Weidemann syndrome which are caused by decreased methylation of the regulatory regions of the UBE3A, and H19 and IGF-2 genes [13]. Alterations to the epigenetic regulation of imprinted genes produce dramatic alterations to the phenotype of the offspring which are evident in early life and so contrast with the phenotypes induced by variations in maternal nutrition throughout pregnancy which become clinically apparent in later life [3].

Differences in the maternal intake of micronutrients during pregnancy in the agouti mouse have been shown to induce differences in the coat colour of the offspring. The murine A^{vy} mutation results from the insertion of an IAP retrotransposon upstream of the agouti gene, which regulates the production of yellow fur pigment. Supplementation of the NIH-31 diet fed to pregnant mice with methyl donors and co-factors such as betaine, choline, folic acid and vitamin B_{12} shifted the distribution of coat colour of the offspring from yellow (agouti) to brown (pseudo-agouti) and increased methylation at the IAP element [14]. Similarly in studies on the Axin Fused ($Axin^{Fu}$) mice, where an IAP insertion into intron 6 of the axin gene causes a kinky tail, supplementation of the maternal diet with methyl donors has also been shown to reduce the incidence of tail kinks in $Axin^{Fu/+}$ offspring by inducing hypermethylation of the IAP element [14]. These findings demonstrate that the level of methylation at these IAP retrotransposons can be modulated by the level of methyl donors within the maternal diet.

Maternal Protein Restriction and Altered Epigenetic Gene Regulation

Feeding a PR diet to pregnant rats is a well-established model of phenotype induction in the offspring, which exhibit some of the characteristics of the metabolic syndrome in humans [2]. Feeding a PR diet to rats during pregnancy induced hypomethylation of the PPARα and GR promoters and increased expression of the GR and PPARα in the liver of the recently-weaned offspring [5]. Hypomethylation of the GR observed in PR offspring was associated with an increase in histone modifications at the GR promoter which facilitate transcription, i.e. acetylation of histones H3 and H4 and methylation of histone H3 at lysine K4, while those that suppress gene expression were reduced or unchanged [15]. Thus, this study showed for the first time that, in contrast to modifying the maternal intake of nutrients associated with 1-carbon metabolism [14], stable changes to the epigenetic regulation of the expression of transcription factors, and hence a phenotype, can be induced in the offspring by modest changes to maternal intake of a macronutrient during pregnancy. Increased expression of the transcription factors PPARα and GR was also accompanied by an increase in expression of their target genes, acyl-CoA oxidase and phosphoenolpyruvate carboxykinase, respectively, which supports the suggestion that such altered epigenetic regulation of transcription factors modifies the activities of important metabolic pathways [5, 15]. Sodium bisulfite treatment followed by pyrosequencing showed that the change in the methylation status of the PPARα promoter was due to hypomethylation of specific CpG dinucleotides, and that the pattern of CpG methylation in day 34 juveniles (fig. 2) persisted in day 80 rats [16]. The altered methylation of specific CpG dinucleotides which correspond to transcription factor binding sites suggests a mechanism by which changes in the epigenetic regulation of genes established during development determines changes in transcription responses to specific stimuli, and thus the capacity of the tissue to respond to metabolic challenge. Thus, altered gene methylation may provide a causal mechanism to explain how maternal diet can induce stable changes in gene expression and phenotype in the offspring.

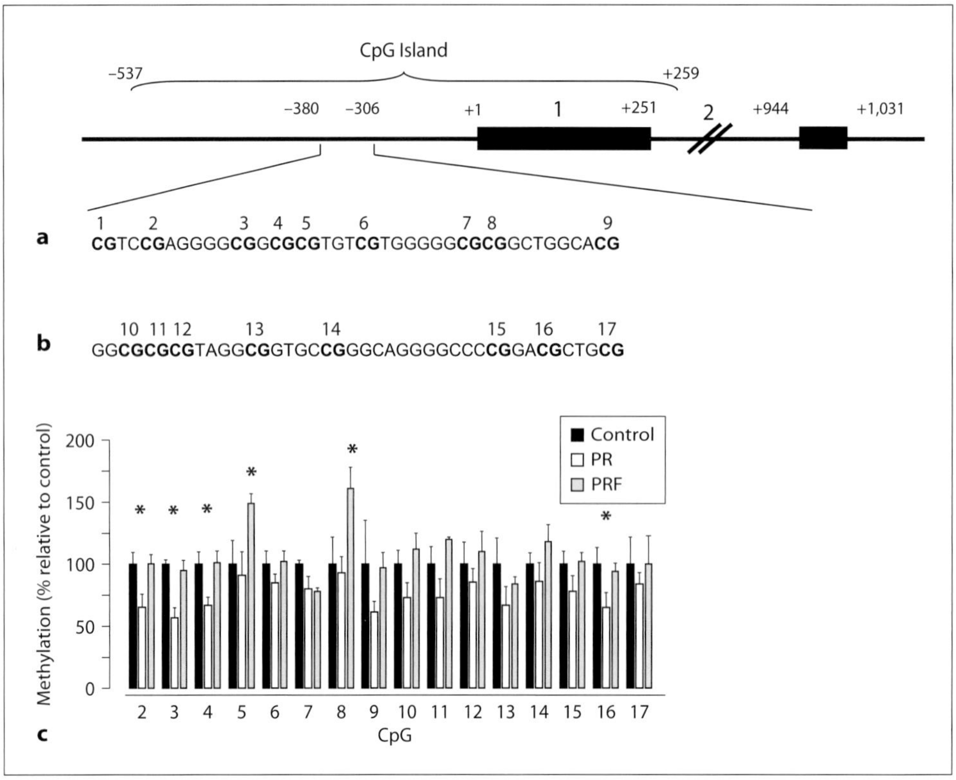

Fig. 2. **a** Structure of the PPARα gene. **b** Nucleotide sequence of the CpG island showing individual CpG dinucleotides. **c** Methylation of individual CpG dinucleotides in the PPARα promoter in the liver of the offspring of rats fed either a control, PR or PR with increased folic acid content (PRF) diet during pregnancy. Data are expressed as mean (SEM). * $p < 0.05$, significantly different from the control group by one-way ANOVA with Dunnett's post-hoc test.

Interestingly, the effects of maternal diet on gene methylation are tissue specific. Maternal protein restriction induces the hypomethylation of PPARα in the liver, heart, umbilical cord and brain but no effect on methylation was been observed in skeletal muscle or adipose tissue (fig. 3). It is not clear how tissue-specific differences in the epigenetic regulation of genes are induced. A difference in embryonic origin cannot completely account for differences in the effect of maternal diet on promoter methylation as hypomethylation of the PPARα and $GR1_{10}$ promoters is reduced in endodermal tissues such as heart and umbilical cord in the offspring of PR dams but not in adipose or skeletal muscle (fig. 3). Alternatively, the binding of transcription factors to promoters in response to an environmental stimulus has been shown to induce specific changes in the epigenetic regulation of genes. Thus, it is possible that the inductive interaction between maternal diet and tissue-specific changes in promoter methylation may depend upon the expression of appropriate transcription factors at the time of the environmental challenge.

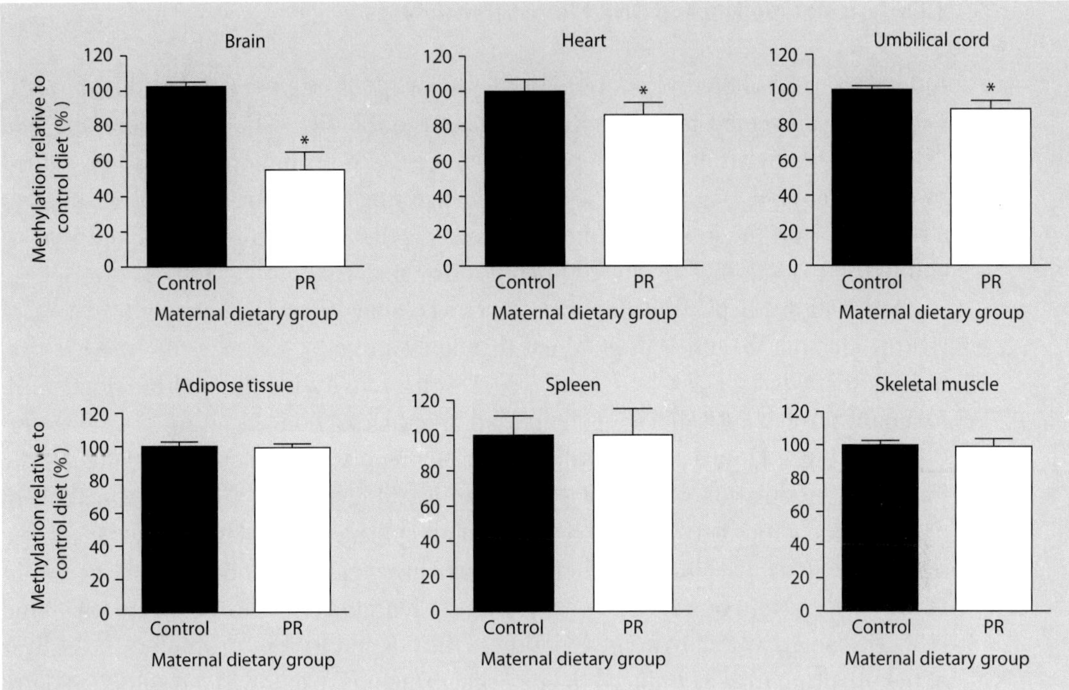

Fig. 3. Maternal protein restriction during pregnancy induces tissue-specific changes in the methylation of the PPARα promoter. Maternal dietary groups were control, 18% (w/w) protein, and PR, 9% (w/w) protein. Data are expressed as mean ± SD. * $p < 0.05$. Significant differences from the control group by one-way ANOVA using Dunnett's post-hoc test.

Transgenerational Transmission of Altered Epigenetic Gene Regulation

Human studies have provided evidence for non-genomic transmission between generations of induced phenotypic traits associated with impaired metabolic homeostasis. Diabetes mortality was increased in men if the paternal grandfather was exposed to abundant nutrition during puberty [17]. During the Dutch Hunger Winter in 1944/5, the daughters of women exposed to nutrient restriction and psychological stress during pregnancy showed decreased birthweight and increased risk of insulin resistance, while their daughters were also born with a lower birthweight [18]. In rats, feeding a PR diet during pregnancy in the F_0 generation resulted in elevated blood pressure and endothelial dysfunction [19] in both the female lineage F_1 and F_2 generations, despite normal nutrition during pregnancy in the F_1 generation. Hypomethylation of the hepatic GR and PPARα promoters was also observed in both the F1 and F2 offspring [20]. However, to date this effect has not been shown in male lineage offspring, or in female lineage to F3, so a truly transgenerational epigenetic effect through meiosis has yet to be demonstrated.

1-Carbon Metabolism and DNA Methyl Transferases

Induction of an altered phenotype in the offspring of rats fed PR diet during pregnancy was prevented by supplementation of the PR diet with glycine or folic acid [21]. Hypomethylation of the hepatic GR and PPARα promoters was also prevented by addition of folic acid to the PR diet [5], suggesting that 1-carbon metabolism plays a central role in the induction of the altered epigenetic regulation of GR and PPARα and in the induction of an altered phenotype by maternal protein restriction.

Methylation of CpG dinucleotides de novo is catalysed by DNA methyltransferases (Dnmt) 3a and 3b, and is maintained through mitosis by gene-specific methylation of hemimethylated DNA by Dnmt1 [6]. Dnmt1 activity is inhibited by Hcyst [22]. Pregnant rats fed a PR diet show increased blood Hcyst concentration in early gestation [23]. Since Dnmt1 expression is negatively regulated by Hcyst and increased by folic acid, modulation of Dnmt1 expression by differences in 1-carbon metabolism may provide a link between maternal diet and epigenetic regulation of gene expression in the fetus. Feeding a PR diet to rats during pregnancy induced a reduction in Dnmt1 expression; however, the expression of Dnmt3a and Dnmt3b were unaltered [15]. This suggests that hypomethylation of the GR and PPARα promoters in the liver of the offspring may be induced by maternal protein restriction as a result of a failure to maintain methylation patterns during mitosis. This is supported by the finding that lower Dnmt1 expression induced by the PR diet was prevented by increasing the folic acid content of the PR diet [15] and is consistent with a central role for Dnmt1 in the induction of an altered phenotype. Loss of Dnmt1 might be expected to result in global demethylation. However, loss of Dnmt1 leads only to a subset of genes being demethylated [24]. Thus, reduced Dnmt1 expression is consistent with hypomethylation of specific genes in the liver in the PR offspring [5].

Observations in Humans

Inevitably, there is considerably less evidence for a role of epigenetics in humans than in animals. However, variations in GR methylation have been reported in human umbilical cord samples. Examination of the methylation status of the GR promoter showed for the first time that among individuals within the normal birth weight range there is considerable variation in the methylation status of the GR gene expressed in this human fetal tissue. Interestingly, Dnmt1 expression predicted 49% of the variation in methylation of the GR promoter expressed in human umbilical cord, suggesting that methylation of the GR promoter in human umbilical cord is associated with the capacity of Dnmt1 to maintain methylation of CpG dinucleotides [20]. Moreover, these findings are consistent with the observations in the rat and mouse, suggesting that induction of different phenotypes in humans by prenatal nutrition may involve variations in (or control of) Dnmt1 expression and, in turn, DNA methylation.

Conclusion

Traditionally, DNA sequence was believed to be the sole determinant of phenotype and phenotypic variation as a result of genetic mutation or recombination. However, there is now increasing evidence to show that epigenetic mechanisms allow a wide range of phenotypes to be generated from one genotype. Moreover, as recent research shows, there are critical periods during development when these epigenetic processes are susceptible to perturbations in maternal nutrition. Understanding the mechanisms responsible for such induced epigenetic changes may provide novel biomarkers of risk and new opportunities for therapeutic strategies to prevent or ameliorate the effects of adverse events in the early life environment on disease risk.

Acknowledgement

G.C.B. and M.A.H. are supported by The British Heart Foundation.

References

1 Godfrey KM, Barker DJ: Fetal programming and adult health. Public Health Nutr 2001;4:611–624.
2 Bertram CE, Hanson MA: Animal models and programming of the metabolic syndrome. Br Med Bull 2001;60:103–121.
3 Gluckman PD, Hanson MA: The developmental origins of the metabolic syndrome. Trends Endocrinol Metab 2004;15:183–187.
4 Burdge GC, Phillips ES, Dunn RL, Jackson AA, Lillycrop KA: Effect of reduced maternal protein consumption during pregnancy in the rat on plasma lipid concentrations and expression of peroxisomal proliferator-activated receptors in the liver and adipose tissue of the offspring. Nutr Res 2004;24:639–646.
5 Lillycrop KA, Phillips ES, Jackson AA, Hanson MA, Burdge GC: Dietary protein restriction of pregnant rats induces and folic acid supplementation prevents epigenetic modification of hepatic gene expression in the offspring. J Nutr 2005;135:1382–1386.
6 Bird A: DNA methylation patterns and epigenetic memory. Genes Dev 2002;16:6–21.
7 Fuks F, Hurd PJ, Wolf D, Nan X, Bird AP, Kouzarides T: The methyl-CpG-binding protein MeCP2 links DNA methylation to histone methylation. J Biol Chem 2003;278:4035–4040.
8 Turner BM: Histone acetylation and an epigenetic code. Bioessays 2002;22:836–845.
9 Gidekel S, Bergman Y: A unique developmental pattern of Oct-3/4 DNA methylation is controlled by a cis-demodification element. J Biol Chem 2002;277: 34521–34530.
10 Hershko AY, Kafri T, Fainsod A, Razin A: Methylation of HoxA5 and HoxB5 and its relevance to expression during mouse development. Gene 2003; 302:65–72.
11 Pham TD, MacLennan NK, Chiu CT, Laksana GS, Hsu JL, Lane RH: Uteroplacental insufficiency increases apoptosis and alters p53 gene methylation in the full-term IUGR rat kidney. Am J Physiol Regul Integr Comp Physiol 2003;285:R962–R970.
12 Weaver IC, Cervoni N, Champagne FA, D'Alessio AC, Sharma S, Seckl JR, Dymov S, Szyf M, Meaney MJ: Epigenetic programming by maternal behavior. Nat Neurosci 2004;7:847–854.
13 Paoloni-Giacobino A: Implications of reproductive technologies for birth and developmental outcomes: imprinting defects and beyond. Expert Rev Mol Med 2006;8:1–14.
14 Waterland RA, Michels KB: Epigenetic epidemiology of the developmental origins hypothesis. Ann Rev Nutr 2007;27:363–388.

15 Lillycrop KA, Slater-Jefferies JL, Hanson MA, Godfrey KM, Jackson AA, Burdge GC: Induction of altered epigenetic regulation of the hepatic glucocorticoid receptor in the offspring of rats fed a protein-restricted diet during pregnancy suggests that reduced DNA methyltransferase-1 expression is involved in impaired DNA methylation and changes in histone modifications. Br J Nutr 2007;97:1064–1073.
16 Lillycrop KA, Phillips ES, Torrens C, Hanson MA, Jackson AA, Burdge GC: Feeding pregnant rats a protein-restricted diet persistently alters the methylation of specific cytosines in the hepatic PPAR alpha promoter of the offspring. Br J Nutr 2008; 100:278–282.
17 Pembrey ME, Bygren LO, Kaati G, Edvinsson S, Northstone K, Sjostrom M, Golding J, ALSPAC Study Team: Sex-specific, male-line transgenerational responses in humans. Eur J Hum Genet 2006;14:159–166.
18 Painter RC, Roseboom TJ, Bleker OP: Prenatal exposure to the Dutch famine and disease in later life: an overview. Reprod Toxicol 2005;20:345–352.
19 Torrens C, Brawley L, Dance CS, Itoh S, Poston L, Hanson MA: First evidence for transgenerational vascular programming in the rat protein restriction model. J Physiol 2002;543:41P–42P.
20 Burdge GC, Slater-Jefferies J, Torrens C, Phillips ES, Hanson MA, Lillycrop KA: Dietary protein restriction of pregnant rats in the F0 generation induces altered methylation of hepatic gene promoters in the adult male offspring in the F1 and F2 generations. Br J Nutr 2007;97:435–439.
21 Jackson AA, Dunn RL, Marchand MC, Langley-Evans SC: Increased systolic blood pressure in rats induced by a maternal low-protein diet is reversed by dietary supplementation with glycine. Clin Sci (Lond) 2002;103:633–639.
22 James SJ, Melnyk S, Pogribna M, Pogribny IP, Caudill MA: Elevation in S-adenosylhomocysteine and DNA hypomethylation: potential epigenetic mechanism for homocysteine-related pathology. J Nutr 2002;132(suppl 8):2361S–2366S.
23 Petrie L, Duthie SJ, Rees WD, McConnell JM: Serum concentrations of homocysteine are elevated during early pregnancy in rodent models of fetal programming. Br J Nutr 2002;88:471–477.
24 Jackson-Grusby L, Beard C, Possemato R, Tudor M, Fambrough D, Csankovszki G, Dausman J, Lee P, Wilson C, Lander E, Jaenisch R: Loss of genomic methylation causes p53-dependent apoptosis and epigenetic deregulation. Nat Genet 2001;27:31–39.

Dr. Karen A. Lillycrop
Cell and Development, School of Biological Sciences
University of Southampton
Bassett Crescent East, Southampton SO16 7PX (UK)
Tel. +44 2380 592948, Fax +44 2380 594459, E-Mail kal@soton.ac.uk

Newnham JP, Ross MG (eds): Early Life Origins of Human Health and Disease.
Basel, Karger, 2009, pp 21–28

The Economics of Developmental Origins of Health and Disease: Modelling the Benefit of a Healthy Start to Life

K.L. Franko[a] · K.C. O'Connor[a] · S.M.B. Morton[a,b]

[a]Liggins Institute, [b]School of Population Health, The University of Auckland, Auckland, New Zealand

The physiological principles underlying the developmental origins of health and disease (DOHaD) hypothesis have been extensively explored and debated and the hypothesis is now widely, although not universally, supported within the realm of biomedical research. Whilst further experimentation is essential for elucidating underlying biological mechanisms, broadening the research beyond the scientific realm is also vital so that the whole body of knowledge produced is applicable outside of scientific laboratories to the world's population. Specifically, determination of the economic burden of poor health outcomes associated with suboptimal intrauterine development makes the DOHaD phenomenon more readily translated into health policy at local, national and international levels. Furthermore, economic analysis of the different life course trajectories that may originate in utero will provide robust evidence upon which to design cost-effective early life health interventions.

It has long been understood that poor birth outcomes across the whole population, particularly rates of LBW, are associated with greater rates of morbidity throughout life from infancy to adulthood [1]. Quantification of actual rather than stylized economic costs associated with a poor birth outcome throughout the life course is needed, followed by description of how these costs impact individual human lives and constrain national economic development. In demonstrating that healthier starts are a means by which nations can achieve economic and development goals, the original promise of early life experience becoming a paradigm shift [2] in the way health policy is designed and delivered may be realized.

The DOHaD hypothesis contends that adverse physiological or morphological states arise in response to impaired environmental conditions during intrauterine development [3], and may be compounded by a later 'mismatch' between the intrauterine and postnatal environments [4]. As maternal poverty during pregnancy increases the likelihood of LBW [5], poverty itself is both a cause and a consequence of economic and physiological adversity throughout life. In particular, economic constraints which impair maternal nutrition, increase maternal work load or increase the likelihood of exposure to infection during pregnancy have all been shown to affect birth outcome negatively [6]. When compared with persons who suffer poor health, adults who are largely free from chronic disease consume relatively fewer medical and social resources, maintain greater productivity and earnings capacity, and may have superior cognitive development [5]. Furthermore, the intergenerational transmission of deprivation may be indirect by poor health and development in childhood increasing the likelihood of ill health in adulthood, and related to poor economic circumstances due to the negative impact of suboptimal development on cognition, school achievement and labour productivity. Thus, a poor start hinders economic opportunity in adulthood, placing that individual's family in poor economic circumstances and creating a cycle of poverty across generations [6]. Under these conditions, deprivation and inequality in households becomes perpetual, reinforced in each generation by the early life environment [7]. Reducing the risk of poor health throughout childhood and adulthood attributable to suboptimal intrauterine development may increase human capital value throughout the life course in all populations across the globe.

Targeting the Life Course

A Starting Point: The Millennium Development Goals

An early attempt at securing a place for the developmental and early life periods in the international spotlight and on health agendas was the production of the Millennium Development Goals (MDGs). Declared with great fanfare nearly two decades ago, the MDGs were intended to raise living standards and improve quality of life across the world [8]. They specifically name reducing the under-5 mortality rate by two thirds between 1990 and 2015 (goal 4) and in the same timeframe, reducing the maternal mortality ratio by three quarters (goal 5 target 1) [9]. Unfortunately, at current rates of progress, neither of these goals will be met by any nation [10]. Predictably, poor countries have fared the worst [10].

A Life History Approach to Health

Despite the MDGs and scientific evidence that early investments in health have the greatest likelihood of creating positive outcomes, a life history approach to health has not gained traction in government or within policy making bodies. Cumulative health throughout development influences the ability of a woman to maintain a healthy pregnancy and provide a healthy start for her offspring, and is influenced by her own periconceptual and gestational health and nutrition [11]. While specific interventions limited to the pregnancy period are likely to have immediate benefits for both mother and offspring, ensuring optimal outcomes also requires investment in health at periods remote from pregnancy itself. As ova are formed in utero, the development of each individual is influenced by the actual environment in the first trimester of his or her maternal grandmother's pregnancy [11]. Growth throughout childhood and adolescence is also strongly influenced by health during infancy and early childhood. Empowering women may delay the age of first pregnancy, and thus increases the likelihood that a woman will be fully grown when she conceives which improves pregnancy outcome. Additionally, maternal height, weight [12] and nutritional status all influence her health status at the time of conception, which in turn promotes the growth and development of her offspring. Optimal nutritional status is imperative throughout pregnancy to ensure appropriate growth and maturity of the fetus at delivery. Context specific investment and intervention strategies during these times will positively influence pregnancy outcome and subsequently, health throughout the life course [11].

Elements of a Poor Start

Critical to promoting health throughout life is an explicit understanding of what constitutes poor health at the start of life. While birth weight is a convenient indicator of intrauterine growth, it alone cannot be used to predict later life health as it does not indicate whether growth is appropriate for the length of gestation, nor does it indicate epigenetic effects or the effects of perturbations that affect development but not growth. While indices of fetal development must include birth weight, a comprehensive list also includes maternal size at conception, gestation length and measures of infant growth and morbidity [11]. Differences in birth weight within and between populations may have different origins, therefore determining relative contributing factors influencing intrauterine growth is essential [11].

Consequences of a Poor Start

There is now sound epidemiological, biological and epigenetic data demonstrating the link between intrauterine and early life health conditions and later health outcomes [13]. Public health priorities tend to focus on birth weight deficits, but both ends of the normal birth weight spectrum are associated with compromised health relative to those in the middle of the range, and there is no birth weight at which an individual is decisively either 'at risk' or 'safe.' Additionally, there are many perturbations which influence life course health and development without necessarily affecting birth weight.

The specific consequences of impaired intrauterine and early life health and the relative importance of specific maternal morbidities, neonatal morbidities, altered growth and predisposition to chronic disease vary across populations according to the stage of economic development and nutrition transition. Nations that have not yet undergone the nutritional transition that accompanies economic growth are frequently impoverished, and malnutrition and the associated depletion of human capital constrain economic growth in a vicious cycle. Early nutritional deficiencies negatively impact growth [14], muscle development [15], cognition [16] and predispose the population to chronic disease [17], all of which convene to limit severely the labour productivity upon which economic growth depends. Similar issues are faced by middle income nations, whose precarious financial situations are further compounded by population health crises, particularly the frequently seen dual burden of undernutrition of youth and overnutrition, and subsequent obesity, amongst adults [18]. Post-nutritional transition states are economically advanced and the nutritional deficiencies and widespread, dire physiological ramifications of poverty are rarely seen. Rather, in wealthy nations, a major focus of health care is cost-effectiveness associated with reducing expenditure [19] and stemming the tide of chronic disease [20].

Life Course Health and Economic Valuation

Ignoring the developmental transfer of biological stock puts future human capital at risk in both health and economic terms [21]. Scant but growing literature suggest a significant economic impact across a lifetime after poor conditions during development [22]. The economic benefits of improved health can be measured both in terms of health care savings and increased productivity [23], although predicting specific economic consequences of a poor start is extremely complex given that feedback loops need to be elucidated, and their economic impacts more fully understood. This level of complexity will require a great deal of cooperation and understanding between epidemiologists and economists to gauge the effectiveness of any particular intervention.

One of the most compelling arguments for prioritizing health in early life is that doing so maximizes the potential for economic productivity and growth [5]. Precisely determining how to optimize economic return on health investment requires a tool to evaluate relative cost-benefit of interventions at different points in the life cycle.

Global Applicability

New early life health interventions will acknowledge the vital importance of pregnancy and fetal development to adult health and well-being, and recognize that birth and life course outcomes are not exclusively biological in origin. Instead, refined health strategies will acknowledge that social circumstance and economic conditions interact in pregnancy with hereditary predisposition and biological environments to determine birth outcome, and that these economic-biologic interactions occurring during the gestational period have consequences that last throughout life and even impact upon the next generation [24]. They also imply a quantifiable economic burden to individuals and the state. Measuring the efficacy of new health strategies requires quantification of the reduced draw on national health and other economic resources such as unemployment and social support.

In rethinking health strategies to emphasize early life, the fact that the burden of a poor birth outcome is not carried solely by the individual and his or her immediate household must be evident. Although it is relatively straightforward to show that LBW infants grow into children and adults who bear larger direct health costs throughout childhood than non-LBW infants [25], it is also possible to examine how variations in the health and economic performance of states are correlated significantly with variation in birth outcome of populations [26]. Even in the wealthier developed nations, measures of child wellbeing and rates of LBW are significantly and inversely associated with measures such as per capita income [27]. The costs that fall on the state such as higher medical supply needs and limits to productive capacity of the labour force further strain a nation's economic resources. Investigations of these costs to date have primarily unveiled the direct resource costs and cost effectiveness of neonatal intensive care, longer hospital admissions and more frequent general practitioner visits for the LBW neonate and infant [28–30]. There is some evidence of the state's indirect costs such as special education needs [31], but additional evidence of other indirect costs, such as lost earnings, forced retirement and the value of reduced national productive capacity, will further support the overall strategic objective of encouraging health and social policy to pay greater attention to early life.

Despite the somewhat limited perspective offered by the current empirical evidence of medical resource costs during early life and childhood of those who were born LBW, it is apparent that the economic load which accompanies the adverse sequelae of a poor birth outcome is a considerable burden on individuals, families and a state's finite resources. It is not likely, however, that the cost burden of an unhealthy start to life is limited to medical expenses, nor is it fully discharged by the end of childhood [32]. The reduced draw on the state's health resources for the care of morbidity in infancy and childhood will be a positive and short-term return to strategies that encourage a healthier birth cohort with a reduced rate of LBW.

Valuing Health

The conceptual challenges in valuing the importance of early life are not limited to the increased risk of mortality and morbidity in early childhood, but continue through each life stage. Analyzing these risks requires standardized burden of disease estimations to quantify the economic effect of the greater prevalence and earlier onset of chronic diseases amongst those in the population who experienced a poor start. Of greater methodological complexity is quantification of the impact of a poor start to life on labour outcomes such as productivity and earnings capacity. That impact is moderated by the functional relationships between poor fetal growth and intellectual performance and ability [33–35]. Since cognitive ability is strongly correlated to educational attainment [36], and since school attainment is directly related to wages [37, 38], then the potential labour market return to a better birth outcome is an increase in earned wage over an entire working life. A similar relationship exists between the prenatal environment, growth trajectories and ultimate adult height [39], and as height and earned income are directly related, implies a substantial impact of a suboptimal early environment on earnings and productivity, especially in underdeveloped nations where manual labour dominates the job market [40, 41]. Additionally, since fetal undernutrition is a risk factor for obesity in later life [42], the model must also account for the economic impact of reduced earnings arising from the portion of obesity in the population with developmental origins.

Challenges Ahead

Inappropriate derivation of causation due to confounding factors including ethnicity, healthcare expenditure, embedded intergenerational attributes and other socioeconomic variables threatens to slow progress toward an early-life focused health sector [43]. Elucidating cause and effect relationships between early life events, health and income inequalities are limited in their explanatory power due to an inability to control experimentally for the above confounders, imprecise definitions of poverty and malnutrition, and characterizations which fail to distinguish relative from absolute poverty [44]. Drawing conclusions about the causal relationships between poverty, deprivation and health at an international level is even more difficult. While there is ample evidence of the lasting adverse effect of income inequality on birth outcomes within individual nations [7, 26, 45], variation between nations in the source and character of income-related data weakens the strength of any global associations [46]. Additionally, due to the time lag between reproductive events, any analysis of intergenerational causation must be cognizant of intervening life events and the pathways that may lead from an event in one generation to the next. Any economic model of such events must disentangle direct from indirect effects of the intrauterine environment, and consider not only the pathway linking biological continuity, but also continuity in the social and physical environment across generations. Only then will we be able to calculate the true costs of a poor start to life.

References

1 UN ACC Standing Committee on Nutrition: Low birthweight: report of a meeting. Dhaka, 2000.

2 Robinson RJ: Is the child father of the man? Controversy about the early origins of cardiovascular disease. BMJ 1992;304:789–790.

3 Barker DJP: The developmental origins of adult disease. J Am Coll Nutr 2004;23(suppl 6):588S–595S.

4 Gluckman PD, Hanson MA: Mismatch: Why Our World No Longer Fits Our Bodies. Oxford, Oxford University Press, 2006.

5 Weil DN: Accounting for the effect of health on economic growth. Q J Econ 2007;122:1265–1306.

6 Spencer N: Weighing the Evidence: How Is Birthweight Determined? San Diego, Radcliffe Medical Press, 2003.

7 Osmani SR: On Inequality; in Blackwell Companion to Sociology. New York, Blackwell, 2001.

8 Investing in development: A practical plan to achieve the Millennium Development Goals. Overview; in UN Millennium Project. New York, United Nations Development Programme, 2005.

9 http://www.un.org/millenniumgoals/. Cited 2009.

10 MDGs and the environment: Agenda for inclusive and sustainable development, in Global monitoring report. The World Bank. Washington, D.C., 2008.

11 World Health Organization: Report of a Technical Consultation: WHO Technical Consultation towards the Development of a Strategy for Promoting Optimal Fetal Development. Geneva, WHO Document Production Services, 2003.

12 Neggers Y, Goldenberg RL: Some thoughts on body mass index, micronutrient intakes and pregnancy outcome 1. Am Soc Nutrition 2003;133:1737s–1740s.

13 Gluckman PD, Hanson MA (eds): Developmental Origins of Health and Disease. Cambridge, Cambridge University Press, 2006.

14 Golden MHN: The nature of nutritional deficiency in relation to growth failure and poverty. Acta Paediatrica 1991;80:95–110.

15 Cheek DB, Hill DE, Cordano A, Graham GG: Malnutrition in Infancy: Changes in muscle and adipose tissue before and after rehabilitation. Pediatr Res 1970;4:135–144.

16 Scrimshaw NS: Malnutrition, brain development, learning, and behavior. Nutrition Research 1998;18:351–379.

17 Barker DJP: Fetal and infant origins of adult disease. Monatsschrift Kinderheilkunde 2001;149:2–6.

18 Caballero B: A nutrition paradox-underweight and obesity in developing countries. N Engl J Med 2005; 352:1514–1516.

19 Grosse SD, Teutsch SM, Haddix AC: Lessons from cost-effectiveness research for United States Public Health Policy, 2007.

20 Antipatis VJ, Kumanyika S, Jeffery RW, Morabia A, Ritenbaugh C: Confidence of health professionals in public health approaches to obesity prevention. Int J Obesity 1999;23:1004–1006.

21 Gluckman PD, Hanson, MA, Bateson P, Beedle AS, Anokhin KV, Bhutta ZA, Bougnères P, Chandak GR, Dasgupta P, Davey Smith G, Ellison PT, Forrester T, Gilbert SF, Jablonka E, Kaplan H, Law CM, Prentice AM, Simpson SJ, Uauy R, West-Eberhard MJ: Towards a new developmental synthesis: Adaptive developmental plasticity and human disease. Lancet 2009;373, in press.

22 Alderman H, Behrman JR: Reducing the incidence of low birth weight in low-income countries has substantial economic benefits. World Bank Res Obs 2006,21:25–48.

23 Alderman H, Behrman JR, Hoddinott J: Economic and nutritional analyses offer substantial synergies for understanding human nutrition. J Nutrition 2007; 137:537–544.

24 Conley D, Bennett NG: Birth weight and income: interactions across generations. J Health Soc Behav 2001;42:450–465.

25 Lewit E, Baker L, Corman H, Shiono P: The direct cost of low birth weight. Future Child 1995;5:35–56.

26 Spencer N: The effect of income inequality and macro-level social policy on infant mortality and low birthweight in developed countries: a preliminary systematic review. Child Care Health Dev 2004;30:699–709.

27 Pickett KE, Wilkinson RE: Child wellbeing and income inequality in rich societies: ecological cross section study. BMJ 2007;335:1054–1055.

28 Almond D, Chay KY, Lee D: The costs of low birth weight. Q J Econ 2005;120:1031–1083.

29 Petrou S: The economic consequences of preterm birth during the first 10 years of life. BJOG 2005; 112(suppl 1):10–15.

30 Rogowski J: Measuring the cost of neonatal and perinatal care. Pediatrics 1999;103:329–335.

31 Chaikind S, Corman H: The impact of low birth weight on special education costs. J Health Econ 1991;10:291–311.

32 Alderman H, Behrman JR: Reducing the incidence of low birth weight in low-income countries has substantial economic benefits. World Bank Res Obs 2006;21:25–48.

33 Jefferis BJMH, Power C, Hertzman C: Birth weight, childhood socioeconomic environment, and cognitive development in the 1958 British birth cohort study. BMJ 2002;325:305–312.

34 Behrman JR: The impact of health and nutrition on education. World Bank Res Obs 1996;11:23–37.

35 Richards M, Hardy R, Kuh D, Wadsworth MEJ: Birthweight, postnatal growth and cognitive function in a national UK birth cohort. Int J Epidemiol 2002;31:342–348.
36 Cawley J, Heckman J, Vytlacil E: Three observations on wages and measured cognitive ability. Lab Econ 2001;8:419–442.
37 Maani S: Investing in Minds: The Economics of Higher Education in New Zealand. Wellington, Institute of Policy Studies, 2007.
38 Behrman JR, Rosenzweig MR: Returns to birthweight. Rev Econ Stat 2004;86:586–601.
39 Joubert K, Gyenis G: Prenatal effects of intra-uterine growth retardation on adult height of conscripts from Hungary. HOMO 2003;54:104–112.
40 Thomas D, Strauss J: Health and wages: evidence on men and women in urban Brazil. J Econometr 1997; 77:159–185.
41 Dinda S, Gangopadhyay PK, Chattopadhyay BP, Saiyed HN, Pal M, Bharati P: Height, weight and earnings among coalminers in India. Econ Hum Biol 2006;4:342–350.
42 Caballero B: Obesity as a consequence of undernutrition. J Pediatr 2006;149(suppl 1):S97–S99.
43 Ben-Shlomo Y, Smith GD: Deprivation in infancy or in adult life: which is more important for mortality risk? Lancet 1991;337:530–534.
44 Vagero D, Illsley R: Explaining health inequalities: beyond Black and Barker. Eur Sociol Rev 1995;11: 219–241.
45 Davey Smith G: Income inequality and mortality: Why are they related? BMJ 1996; 312:987.
46 Judge K, Mulligan JA, Benzeval M: Income inequality and population health. Soc Sci Med 1998;46:567–579.

Dr. Kathryn L. Franko
Liggins Institute, The University of Auckland
Private Bag 92019
Auckland 1023 (New Zealand)
Tel. +64 9 373 7599, ext. 83450, Fax +64 9 373 7497, E-Mail k.franko@auckland.ac.nz

Newnham JP, Ross MG (eds): Early Life Origins of Human Health and Disease.
Basel, Karger, 2009, pp 29–41

Nutritional Interventions in Mothers to Improve the Health of the Offspring – Are We Ready?

Caroline H.D. Fall[a] · Jane E. Harding[b] · Chittaranjan S. Yajnik[c]

[a]MRC Epidemiology Resource Centre, University of Southampton, Southampton General Hospital, Southampton, UK; [b]Department of Neonatology, Liggins Institute, The University of Auckland, Auckland, New Zealand; [c]Diabetes Unit, King Edward Memorial Hospital, Pune, India

Over the past 20 years a large body of research has accumulated, showing associations between birthweight and the risk of developing obesity, cardiovascular disease, type 2 diabetes and other health outcomes in later life. This has highlighted the importance of optimal fetal nutrition for lifelong health and has brought a renewal of interest, from policy makers, in improving maternal nutrition. In a 2006 report *'Repositioning nutrition as central to development'* the World Bank recommended that interventions to eradicate undernutrition in low-income countries should focus on pregnant women and children under 2 years of age in order to reap the greatest long-term benefits for human capacity and health [1]. The World Health Organisation highlighted the global burden of death, disability and lost human potential resulting from impaired fetal development [2]. And a series of papers published in the *Lancet* described the large burden of maternal and infant undernutrition in low-income countries and its adverse consequences throughout life [3, 4].

Three new reports in the UK have highlighted problems with maternal nutrition in high-income countries too: the UK government's Foresight report on tackling obesity [5]; the National Institute for Clinical Excellence's review on improving the nutrition of pregnant and breastfeeding mothers [6], and the British Medical Association's report on maternal and infant nutrition [7]. Each report has a different focus, but three key points emerge. Firstly, that even in high-income countries there are concerns about poor-quality diets in young women, secondly that there is an increasing problem of maternal obesity, and thirdly that both of these are likely to have adverse consequences for the long-term health of their children.

The Nutritional State of the World's Mothers

In low-income countries, large numbers of women are underweight, stunted and micronutrient deficient [3] (fig. 1). Maternal undernutrition is a major cause of low birthweight in these countries (both intrauterine growth restriction and pre-term delivery). Countries in rapid economic transition are now experiencing a 'double burden of malnutrition'. On the one hand, diets are still poor in terms of micronutrient and macronutrient quality. For example, in India vitamin B_{12}, vitamin D and iron deficiency are common in women of reproductive age, and intakes of foods containing vitamin A, zinc and calcium are often low [8–10]. On the other hand, especially in cities, overweight and obesity among women are on the rise (fig. 1). This is related to energy-dense diets and increasingly sedentary work patterns; for example the burgeoning information technology and call-centre industries in India employ large numbers of young women, working indoors in extremely sedentary conditions. The complications of obesity, such as gestational diabetes, occur at much lower thresholds of body weight in India than in white Caucasians [11]. At any body weight, Indian women have a higher percentage body fat; this phenomenon may itself be a consequence of intrauterine malnutrition and maladaptation.

In high-income countries, obesity is reaching epidemic proportions in young women (fig. 1). During pregnancy, it increases the risk of pre-eclampsia, gestational diabetes, stillbirth, pre-term delivery, large-for-gestational age births and caesarean section [12]. Poor-quality diets are an issue in developed countries too, especially among disadvantaged groups. In the UK, younger women are more likely to consume 'fast foods', savoury snacks, and fizzy drinks, and less likely to eat wholemeal bread, cereals, fruit and vegetables and oily fish [13]. In Sydney, one third of women giving birth to small-for-gestational age babies were found to have an eating disorder [14]. The UK About Teenage Eating Study [15] has shown that low micronutrient status in young women is common, and associated with poor birth outcomes. Guidelines on pre-conceptional nutrition (e.g. folate supplements to prevent fetal anomalies) are frequently not followed in this vulnerable group, because of unplanned pregnancy.

Evidence Linking Maternal Nutrition in Pregnancy to Long-Term Outcomes in the Offspring

Animal Studies

Numerous animal studies provide convincing evidence that maternal nutrition during pregnancy can have profound effects on the later health of the offspring. In rodents, experimental approaches have included global undernutrition, low-protein and high-fat diets in pregnancy, all of which may lead to altered body composition,

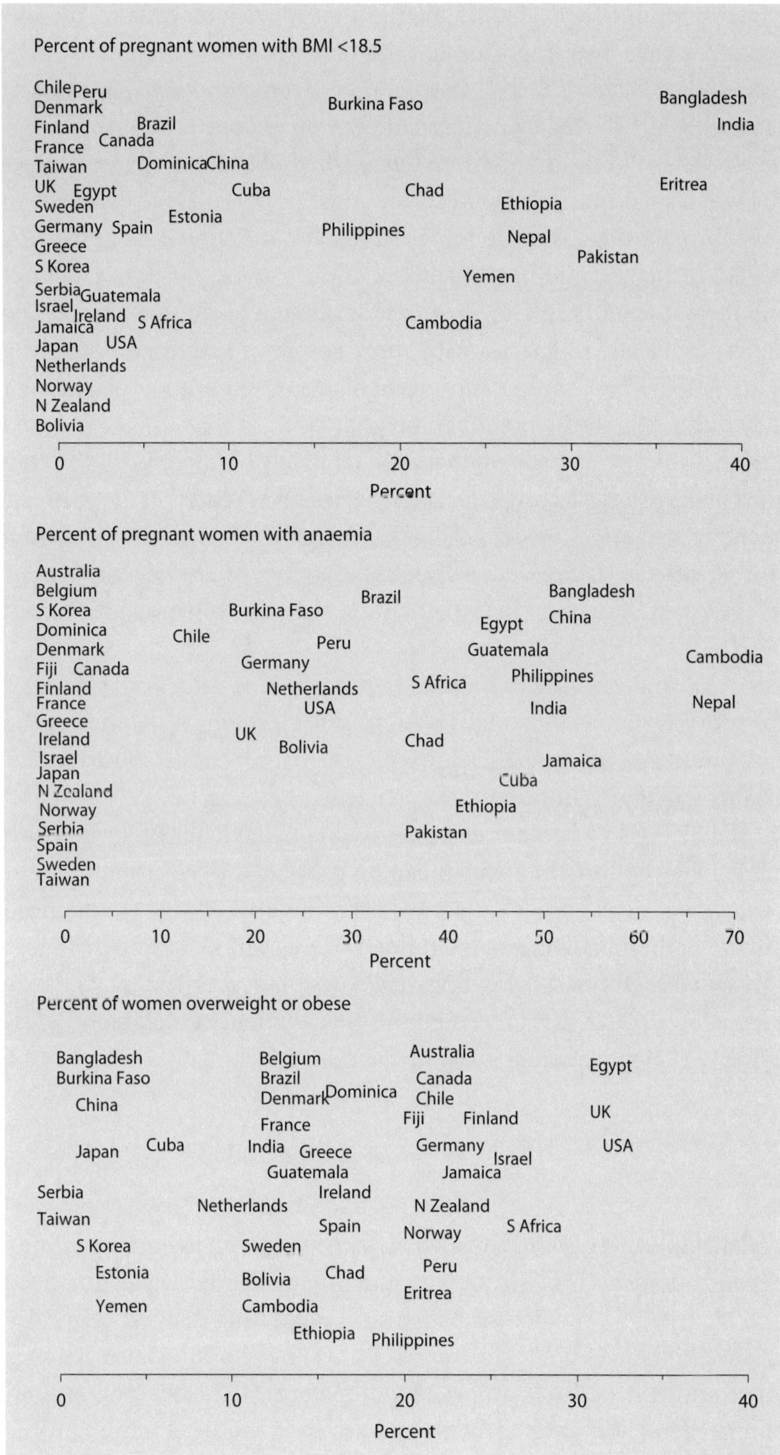

Fig. 1. Approximate percentages of pregnant women with low BMI and anaemia and women with overweight or obesity in various countries.

increased blood pressure, impaired glucose tolerance and dyslipidaemia in the offspring. Similar changes have been reported in other species, including guinea pigs, pigs, horses, sheep and primates [16, 17]. Postnatal fat deposition, activity, appetite and food preferences are all affected by maternal diet in pregnancy [18].

Animal studies have also been used to investigate whether there are critical times during pregnancy when such nutritional effects are most important. For example, in rats, a maternal low-protein diet results in impaired insulin secretion in the offspring if the diet is imposed during the middle or last week of the 3-week pregnancy, but not if imposed during the 1st week [19]. However, care should be taken in extrapolating from animal studies, especially in species with short gestation length and offspring born very immature, as the rate, timing and mechanisms of maturation of different organ systems may differ from those of human pregnancy.

Importantly, animal studies have shown that even relatively brief periods of altered maternal nutrition can have long-lasting effects on the offspring, and that these effects can be independent of any effect on size at birth. In sheep, for example, maternal undernutrition for as little as 10 days in a 5-month pregnancy altered hypothalamo-pituitary-adrenal (HPA) axis function in the offspring, although birthweight was not affected [20].

Animal studies are also beginning to provide the first robust evidence that nutritional interventions during pregnancy may be useful in improving outcomes in the offspring, and to indicate which specific nutrients might be critical. Taurine, for example, is an amino acid that is important for pancreatic beta cell development and insulin secretion. Rats fed a low-protein diet during pregnancy have offspring with impaired pancreatic beta cell function, which can be prevented by provision of taurine in the drinking water of the pregnant dams [21]. Similarly, glycine is an amino acid important in one-carbon metabolism including DNA synthesis and methylation. Folate is an important co-factor in these pathways. Provision of supplemental glycine to pregnant rats fed a low protein diet prevented the development of high blood pressure in the offspring [22]. Supplemental folate had similar effects [23].

Human Studies

Newborn size is related to maternal energy balance, increasing with the mother's body mass index (BMI) and adiposity [24]. Her height, and independently, leg length, head circumference and birthweight, predict the baby's size, suggesting that her own early nutritional experience influences how she nourishes the fetus [24, 25]. David Barker's original fetal programming hypothesis proposed that maternal ill health, poverty and undernutrition impair fetal and infant growth and are root causes of adult chronic disease [26]. If this is so, indices of poor maternal nutrition would be associated with cardiovascular disease or its risk factors in the offspring. So far, unfortunately, there are only crude data available to test this hypothesis.

Studies in low-income or historically poor countries have shown that low maternal weight, BMI or skinfolds in pregnancy are associated with higher offspring blood pressure [27, 28], insulin resistance [29, 30] and risk of coronary heart disease [31]. Follow-up of the 1944–1945 Dutch Hunger Winter showed that exposure of the mother to acute famine was associated with increased cardiovascular risk in the adult offspring. Consistent with the animal studies, outcomes varied according to the timing of famine exposure [32]; late gestation exposure was associated with glucose intolerance and early gestation exposure with obesity, atherogenic lipid profiles and coronary heart disease. Famine effects were independent of birthweight, suggesting that maternal undernutrition may impair adult health without reducing size at birth.

The INCAP trial in Guatemala (1969–1977) is the only nutritional intervention study in which cardiovascular risk factors have been measured in the adult offspring [33]. In the original trial, pregnant women and children received one of two supplements: a high-protein, high energy drink (Atole) or a lower nutrient drink (Fresco). There was no difference in birthweight between these two groups, but the children of mothers who received atole had lower triglyceride and higher HDL cholesterol concentrations in adult life.

The Pune Maternal Nutrition Study

The Pune Maternal Nutrition Study (PMNS) was established specifically to investigate the relationship of maternal nutrition to the children's future risk of diabetes and heart disease and has provided further evidence of nutritional programming of these disorders in fetal life [8, 9, 11, 34]. It collected detailed information on diet, workload and micronutrient status in over 800 pregnant women in 6 rural Indian villages. The mothers weighed on average 42 kg and were 1.52 m tall, and the majority performed heavy domestic and farming work. Mean birthweight was 2.7 kg and the babies were thin compared to UK babies but had comparable subscapular skinfold thickness [11]. The mother's pre-pregnant body size and consumption of green leafy vegetables, fruit and milk (but not protein or energy) during pregnancy were strongly predictive of newborn size [8]. Two thirds of mothers had vitamin B_{12} deficiency but folate deficiency was rare; higher folate and vitamin B_{12} concentrations predicted larger neonatal size, and high homocysteine concentrations predicted intrauterine growth restriction.

At 6 years of age, the children were thin by international standards but had higher percentage body fat than European children. Maternal intake of green leafy vegetables and red cell folate concentrations were independent predictors of the child's adiposity (DXA). Folate concentrations also predicted higher insulin resistance in the child, and children born to mothers with the lowest vitamin B_{12} concentrations and highest folate concentrations were the most insulin resistant [34] (fig. 2). Thus, an imbalance in two methylating vitamins in the mother predicted higher adiposity and insulin resistance in the child. These findings add to evidence from experimental studies in

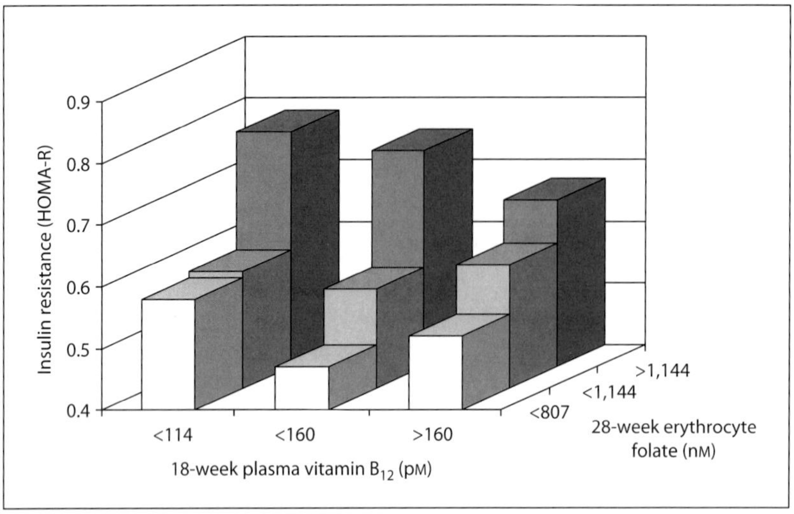

Fig. 2. Insulin resistance (HOMA) in the Pune Maternal Nutrition Study children at 6 years in relation to maternal vitamin B_{12} and erythrocyte folate concentrations in pregnancy. Reproduced from reference [34].

animals suggesting an important role for one-carbon metabolism in the fetal programming of type 2 diabetes [22, 23, 35, 36].

Maternal Obesity and Gestational Diabetes

Recent studies linking birthweight and adult disease have shown that individuals at *both* extremes of birthweight are at increased risk of adult diabetes [37]. It is now well established that men and women who were macrosomic at birth because their mothers were diabetic during pregnancy have increased risks of obesity and diabetes [38]. Initially shown in the Pima Indians, this seems to occur in all populations. The mother-to-offspring transmission of diabetes may have a genetic component, but the main driver is thought to be hyperglycaemia and hyperinsulinism in the fetus of the diabetic mother, an example of 'fuel-mediated teratogenesis' as proposed by Freinkel [39]. Schoolchildren in Taiwan showed a similar phenomenon [40] and Indian children of gestationally diabetic mothers had increased adiposity and hyperinsulinaemia at the age of 5 years [41].

Maternal obesity and being born large for gestational age, even in the absence of gestational diabetes, predict higher metabolic risk factors in childhood [42]. Fuel-mediated teratogenesis may therefore occur even in 'normoglycaemic' mothers. In a recent trial of two diets, normoglycaemic mothers on a high glycaemic index diet gave birth to more large-for-gestational-age babies [43]. Children of mothers with higher glucose concentrations within the normal range are more adipose [44]. With increasing numbers of overweight women, such effects may make an important contribution to the burden of obesity and diabetes in the near future.

Importance of Pre-Pregnant/Peri-Conceptional Nutrition

There is growing evidence that maternal nutritional status before and around the time of conception may be particularly important for the later health of the offspring. In sheep, moderate maternal undernutrition before and for up to 1 month after conception results in altered fetal growth, metabolism, endocrine status and cardiovascular regulation, despite a high plane of maternal nutrition for the rest of the pregnancy [45]. Importantly, fetal HPA axis activation is accelerated, leading in about half the pregnancies to preterm birth [46]. The offspring show altered growth regulation, impaired glucose tolerance and altered HPA axis function in early adult life [47].

Furthermore, the period of peri-conceptional undernutrition required to have long-term effects on the offspring may be very brief. In sheep, maternal undernutrition for as little as 7 days after mating is sufficient to alter fetal HPA axis function [48]. In rats, a low-protein diet fed only during the 4.5-day pre-implantation period results in offspring with elevated blood pressure and impaired glucose tolerance [49].

The mechanisms underlying such long-term effects of brief nutritional insults are not yet clear. However animal experiments suggest that the early embryo develops in an environment of altered nutrient and growth factors, including low levels of cortisol, glucose and amino acids. In turn, these appear to alter the pattern of embryonic growth, including altered cell cycle length and altered distribution of blastocyst cells between the inner and outer cell mass [49]. Maternal metabolic and endocrine adaptation to pregnancy is also affected; peri-conceptional undernutrition in sheep alters uterine blood flow in late pregnancy [50] and prevents the normal development of maternal insulin resistance (fig. 3) [51]. This would in turn reduce the supply of critical nutrients to the fetus, thus altering fetal growth and endocrine regulation. Furthermore, impaired maternal adaptation to pregnancy is seen even if undernutrition is limited to the period before mating, suggesting that maternal nutritional status before pregnancy may itself have important effects on the later health of the offspring.

There is limited evidence for the importance of pre- or peri-conceptional nutritional status in human pregnancy, but similar effects seem likely. A history of pre-pregnancy eating disorders, low body mass at conception, and severe nausea and vomiting in early pregnancy are all associated with altered fetal growth trajectory, and an increased risk of low birthweight and preterm birth [52, 53]. Preterm birth itself is associated with increased blood pressure and impaired glucose tolerance in adulthood [54]. Babies conceived during the Dutch famine, but not those exposed later in pregnancy, were also at increased risk of cardiovascular disease in adult life

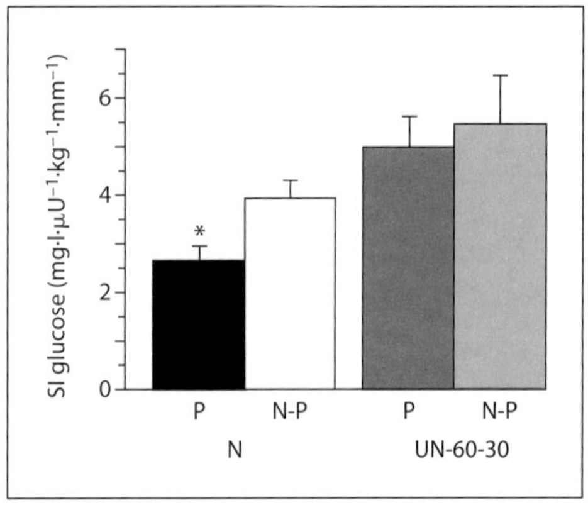

Fig. 3. Maternal insulin sensitivity of glucose metabolism in non-pregnant (NP) and pregnant (P) sheep at 65 days gestation (term = 147 days). Peri-conceptional undernutrition from 60 days before until 30 days after mating (UN-60–30) prevents the normal decrease in insulin sensitivity seen with pregnancy in well-nourished (N) animals (* p = 0.02) [51].

[32]. Thus, nutritional interventions to improve the health of the offspring are likely to have most benefit if started before pregnancy.

Recent Intervention Trials in Human Pregnancy

Many trials have tested the effect on birthweight of maternal nutritional supplementation starting in mid-late pregnancy, and some have shown effects that might be expected to improve long-term outcomes. In the Gambia, a high-energy biscuit for severely undernourished pregnant women increased birthweight by more than 130 g [55] and most other trials of balanced energy and protein supplementation have shown small increases in birthweight (around 40 g overall) and important reductions in small-for-gestational age births [56]. The same is true of multiple micronutrient supplementation trials [57]. Maternal supplementation with fish oil during pregnancy may increase birthweight and reduce the risk of very preterm birth [58], and calcium supplementation reduces the risk of preterm birth in high-risk women [59]. No trials have tested the effect of pre-conceptional supplementation on fetal growth and newborn size, but there are ongoing trials assessing this [60].

Birthweight is a crude measure of fetal development, and may underestimate effects on specific fetal tissues and organs, and thus on functional and long-term outcomes. The Gambia high-energy biscuit trial, and a recent trial of multiple micronutrient supplementation in Indonesian mothers reported significantly reduced infant mortality [55, 61], although opposing results were found in two other multiple micronutrient trials in Nepal [62]. The reasons for these differences are unknown, but may relate to the height of the mother and her nutritional status at the time of conception.

Some trials have examined longer-term outcomes in the children, including blood pressure and growth. In the Gambia trial, blood pressure was no different in the adolescents whose mothers were in the experimental group (supplemented

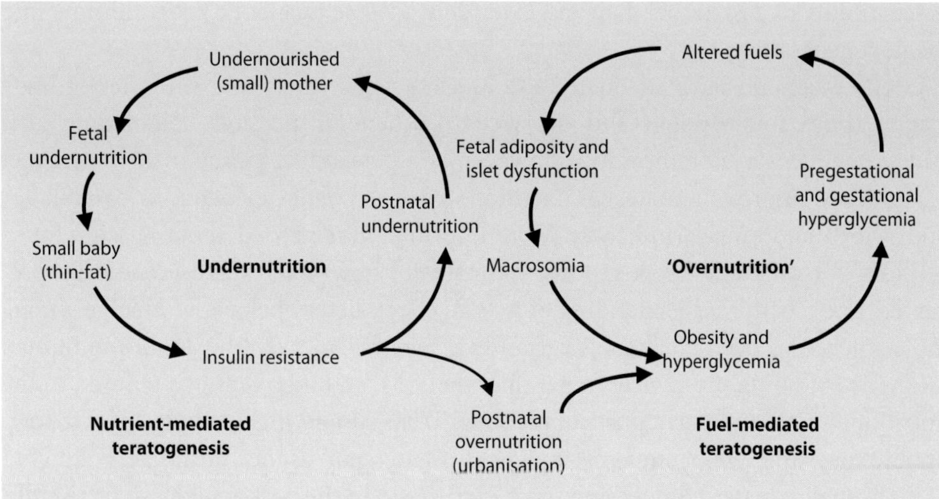

Fig. 4. Fetal programming in rapid transition. Modified from reference [67].

in pregnancy) compared with the control group (supplemented during lactation) [63]. In contrast, children of mothers supplemented with calcium [64] or multiple micronutrients [65] had lower blood pressures. The latter trial (in Nepal) showed sustained effects of maternal multiple micronutrient supplementation on the children's growth; at 2 years, the children had larger head, chest, mid-upper arm and hip circumferences and triceps skinfolds and fewer were underweight and stunted [65]. Further long-term follow-up of randomised trials are required to determine whether these short-term differences translate into longer-term effects on the health of the offspring.

Nutritional Interventions – Are We Ready?

What messages about nutrition can DOHaD research give to policy makers and women preparing for pregnancy? Studies in animals and preliminary data from humans show that a mother's nutrition can have important effects on the future health of the fetus. Pre-conceptional nutrition is probably at least as important as nutrition during pregnancy itself. The nutritional state of many women, in both developing and developed countries, is far from healthy. In high-income countries, poor-quality diets are common, especially among younger women. In low-income countries, while many women remain energy-deficient and have multiple micronutrient deficiencies, economic transition is increasingly causing another form of malnutrition: obesity and gestational diabetes [66, 67]. Both extremes have been linked to ill-health in the offspring in later life (fig. 4). Undernutrition of females in early life increases their susceptibility to gestational diabetes, extending adverse metabolic programming into the next generation.

Mothers should have adequate food and nutrients, and avoid extremes of body weight (both underweight and overweight) whenever possible. Persuading girls and women to eat more healthily, however, and persuading policy makers to direct resources to improving maternal nutrition, will need better evidence of benefits for both short- and long-term health. More information is needed about specific interventions for different target groups, estimates of how much disease could be prevented, and a better understanding of how to effect dietary behaviour change among the women who need it. There is certainly a need for intervention studies in human mothers, including pre-conceptional interventions, and interventions tailored to the nutritional profile of particular populations. Trials should include long-term follow-up, and measurement of outcomes that really matter, not just newborn size. Follow-up of high-quality maternal nutrition trials carried out in the past would also be useful.

Acknowledgements

This work was supported by the Medical Research Council (UK), the Wellcome Trust, the Parthenon Trust, the British Heart Foundation, the Health Research Council of New Zealand, the National Research Centre for Growth and Development, New Zealand, and the Department of Biotechnology, Government of India, New Delhi. Dr. Anne Jaquiery kindly supplied the data for figure 3.

References

1 World Bank: Repositioning nutrition as central to development; a strategy for large-scale action. World Bank, Washington, 2006.

2 World Health Organisation: Promoting Optimal Fetal Development. Report of a Technical Consultation. World Health Organisation, Geneva, 2006; http://www.who.int/nutrition/topics/fetal_dev_report.pdf.

3 Black RE, Allen LH, Bhutta Z, Caulfield LE, de Onis M, Ezzati M, Mathers C, Rivera J, the Maternal and Child Undernutrition Study Group: Maternal and child undernutrition: global and regional exposures and health consequences. Lancet 2008;371:243–260.

4 Victora CG, Adair L, Fall C, Hallal PC, Martorell R, Richter L, Sachdev HPS, the Maternal and Child Undernutrition Study Group: Maternal and child undernutrition: consequences for adult health and human capital. Lancet 2008;371:340–357.

5 Foresight Project. Tackling obesities: Future Choices. UK Government Foresight Programme, 2007. www.foresight.gov.uk/Obesity/obesity_final/17.pdf.

6 Improving the nutrition of pregnant and breastfeeding mothers and children in low-income households. NICE public health guidance no. 11. National Institute for Clinical Excellence 2008. www.nice.org.uk/guidance/ph11/guidance/pdf/.

7 British Medical Association: Early life nutrition and lifelong health. British Medical Association, London, 2009.

8 Rao S, Yajnik CS, Kanade A, Fall CHD, Margetts BM, Jackson AA, Shier R, Joshi S, Rege S, Lubree H, Desai B: Maternal fat intakes and micronutrient status are related to fetal size at birth in rural India; the Pune Maternal Nutrition Study. J Nutr 2001; 131:1217–1224.

9 Ganpule A, Yajnik CS, Fall CHD, Rao S, Fisher DJ, Kanade A, Cooper C, Naik S, Joshi N, Lubree H, Deshpande V, Joglekar C: Bone mass in Indian children; relationships to maternal nutritional status and diet during pregnancy; the Pune Maternal Nutrition Study. J Clin Endocrinol Metab 2006;91: 2994–3001.

10 Farrant HJW, Krishnaveni GV, Hill JC, Boucher BJ, Fisher DJ, Noonan K, Osmond C, Veena SR, Fall CHD: Vitamin D insufficiency is common in Indian mothers but is not associated with gestational diabetes or variation in newborn size. Eur J Clin Nutr 2008; Epub ahead of print.

11 Yajnik CS: Obesity epidemic in India; intrauterine origins? Proc Nutr Soc 2004;63:1–10.

12 Bhattacharya S, Campbell DM, Liston WA, Bhattacharya S: Effect of body mass index on pregnancy outcomes in nulliparous women delivering singleton babies. BMC Public Health 2007;7:168.

13 Office of National Statistics: The National Diet and Nutrition Survey: adults aged 19 to 64 years. Office of National Statistics, London, 2002; www.statistics.gov.uk/ssd/surveys/national_diet_nutrition_survey_adults.asp.

14 Conti J, Abraham S, Taylor A: Eating behavior and pregnancy outcome. J Psych Res 1998;44:465–477.

15 Wheeler SJ, Sanders TAB, Thomas J, Seed PT, Briley AL, Hutchinson C, Baker PN, Poston L: Relationship between diet, nutritional status and adverse pregnancy outcomes in UK adolescent pregnancies; the About Teenage Eating (ATE) Study. Early Hum Dev 2007;83(suppl 1):S170.

16 McMillen I, Robinson JS: Developmental origins of the metabolic syndrome: prediction, plasticity, and programming. Physiol Rev 2005;85:571–633.

17 Fowden AL, Giussani DA, Forhead AJ: Endocrine and metabolic programming during intrauterine development. Early Hum Dev 2005;81:723–734.

18 Langley-Evans SC, Bellinger L, McMullen S: Animal models of programming: early life influences on appetite and feeding behaviour. Mat Child Nutr 2005;1:142–148.

19 Alvarez C, Martin MA, Goya L, Bertin E, Portha B, Pascual-Leone AM: Contrasted impact of maternal rat food restriction on the fetal endocrine pancreas. Endocrinology 1997;138:2267–2273.

20 Bloomfield FH, Oliver MH, Giannoulias CD, Gluckman PD, Harding JE, Challis JRG: Brief undernutrition in late-gestation sheep programs the hypothalamic-pituitary-adrenal axis in adult offspring. Endocrinology 2003;144:2933–2940.

21 Cherif H, Reusens B, Ahn MTh, Hoet J-J, Remacle C: Effects of taurine on the insulin secretion of rat fetal islets from dams fed a low protein diet. J Endocrinol 1998;159:341–348.

22 Jackson AA, Dunn RL, Marchand MC, Langley-Evans SC: Increased systolic blood pressure in rats induced by a maternal low-protein diet is reversed by dietary supplementation with glycine. Clin Sci 2002;103:633–639.

23 Torrens C, Brawley L, Anthony FW, Dance CS, Dunn R, Jackson AA, Poston L, Hanson MA: Folate supplementation during pregnancy improves offspring cardiovascular dysfunction induced by protein restriction. Hypertension 2006;47:982–987.

24 Leary S, Fall CHD, Osmond C, Lovel H, Campbell D, Eriksson J, Forrester T, Godfrey K, Hill J, Jie M, Law C, Newby R, Robinson S, Yajnik CS: Geographical variation in relationships between parental body size and offspring phenotype at birth. Acta Obstet Scand 2006;85:1066–1079.

25 Martin RM, Davey Smith G, Frankel S, Gunnell D: Parent's growth in childhood and the birth weight of their offspring. Epidemiology 2004;15:308–316.

26 Barker DJP: The fetal origins of coronary heart disease. BMJ 1995;311:171–174.

27 Adair LS, Kuzawa BBA, Borja J: Maternal energy stores and diet composition during pregnancy program adolescent blood pressure. Circulation 2001;104:1034–1039.

28 Godfrey KM, Forrester T, Barker DJP, et al: Maternal nutritional status in pregnancy and blood pressure in childhood. Br J Obstet Gynaecol 1994;101:398–403.

29 Jie M, Law C, Zhang K-L, Osmond C, et al: Effects of infant birthweight and maternal body mass index in pregnancy on components of the Insulin Resistance Syndrome in China. Ann Intern Med 2000;132:253–260.

30 Eriksson JG, Forsen T, Tuomilhehto J, Jaddoe VWV, Osmond C, Barker DJP: Effects of size at birth and childhood growth on the insulin resistance syndrome in elderly individuals. Diabetologia 2002;45:342–348.

31 Stein C, Fall CHD, Kumaran K, et al: Fetal growth and coronary heart disease in South India. Lancet 1996;348:1269–1273.

32 Roseboom TJ, van der Meulen J, Ravelli ACJ, Osmond C, Barker DJP, Bleker OP: Effects of prenatal exposure to the Dutch famine on adult disease in later life: an overview. Mol Cell Endocrinol 2001;185:93–98.

33 Stein AD, Wang M, Ramirez-Zea M, Flores R, Grajeda R, Melgar P, Ramakrishnan U, Martorell R: Exposure to a nutrition supplementation intervention in early childhood and risk factors for cardiovascular disease in adulthood: evidence from Guatemala. Am J Epidemiol 2006;164:1160–1170.

34 Yajnik CS, Deshpande SS, Jackson AA, Refsum H, Rao S, Fisher DJ, Bhat DS, Naik SS, Coyaji KJ, Joglekar CV, Joshi N, Lubree HG, Deshpande VU, Rege SS, Fall CHD: Vitamin B12 and folate concentrations during pregnancy and insulin resistance in the offspring: The Pune Maternal Nutrition Study. Diabetologia 2008;51:29–38.

35 Sinclair KD, Allegruci C, Singh R, Gardner DS, Sebastian S, Bispham J, Thurston J, Huntley JF, Rees WD, Maloney CA, Lea RG, Craigon J, McEvoy TG, Young LE: DNA methylation, insulin resistance, and blood pressure in offspring determined by maternal periconceptional B vitamin and methionine status. Proc Natl Acad Sci 2007;104:19351–19356.
36 Waterland RA, Jirtle RL: Transposable elements: targets for early nutritional effects on epigenetic gene regulation. Mol Cell Biol 2003;23:5293–5300.
37 Rich-Edwards JW, Colditz GA, Stampfer MJ, Willett WC, Gillman MW, Hennekens CH, et al: Birthweight and the risk of type 2 diabetes in adult women. Ann Int Med 1999;130:278–284.
38 Dabelea D, Hanson RL, Lindsay RS, Pettitt DJ, Imperatore G, Gabir MM, Roumain J, Bennett PH, Knowler WC: Intrauterine exposure to diabetes conveys risks for type 2 diabetes and obesity; a study of discordant sibships. Diabetes 2000;49:2208–2211.
39 Freinkel N: Of pregnancy and progeny. Diabetes 1980;29:1023–1035.
40 Wei J-N, Sung F-C, Li C-Y, Chang C-H, Lin R-S, Lin C-C, et al: Low birthweight and high birthweight infants are both at risk to have type 2 diabetes among schoolchildren in Taiwan. Diabetes Care 2003;26: 343–348.
41 Krishnaveni GV, Hill JC, Leary S, Veena SR, Saperia J, Saroja A, Karat SC, Fall CHD: Anthropometry, glucose tolerance and insulin concentrations in Indian children: relationships to maternal glucose and insulin concentrations during pregnancy. Diabetes Care 2005;28:2919–2925.
42 Boney CM, Verma A, Tucker R, Vohr BR: Metabolic syndrome in childhood: association with birth weight, maternal obesity, and gestational diabetes mellitus. Pediatrics 2005;115:e290–e296.
43 Moses RG, Luebcke M, Davis WS, Coleman KJ, Tapsell LC, Petocz P, Brand-Miller JC: Effect of a low-glycemic-index diet during pregnancy on obstetric outcomes. Am J Clin Nutr 2006;84:807–812.
44 Hillier TA, Pedula KL, Schmidt MM, Mullen JA, Charles MA, Pettitt DJ: Childhood obesity and metabolic imprinting: the ongoing effects of maternal hyperglycemia. Diabetes Care 2007;30:2287–2292.
45 Bloomfield FH, Oliver MH, Harding JE: The late effects of fetal growth patterns. Arch Dis Child Fetal Neonatal Ed 2006;91:F299–F304.
46 Bloomfield FH, Oliver MH, Hawkins P, Campbell M, Phillips DJ, Breier BH, Gluckman PD, Challis JRG, Harding JE: A periconceptual nutritional origin for non-infectious preterm birth. Science 2003; 300:606.
47 Todd SE, Oliver MH, Jaquiery AL, Bloomfield FH, Harding JE: Periconceptional undernutrition of ewes decreases glucose tolerance in their postnatal offspring. Early Human Dev 2007;83(suppl 1):S53.
48 MacLaughlin SM, McMillen IC: Impact of periconceptional undernutrition on the development of the hypothalamo-pituitary-adrenal axis: does the timing of parturition start at conception? Current Drug Targets 2007;8:880–887.
49 Fleming TP, Kwong WY, Porter R, Ursell E, Fesenko I, Wilkins A, Miller DJ, Watkins AJ, Eckert JJ: The embryo and its future. Biol Reprod 2004;71:1046–1054.
50 Rumball CWH, Bloomfield FH, Harding JE: Cardiovascular adaptations to pregnancy in sheep and effects of periconceptional undernutrition. Placenta 2008;29:89–94.
51 Jaquiery AL, Oliver MH, Rumball C, Buckley AJ, Harding JE: Effect of periconceptional undernutrition on insulin sensitivity at 65 and 120 days gestation in singleton bearing pregnant ewes. Pediatr Res 2005;58:P3–053:1114.
52 Sollid CP, Wisborg K, Hjort J, Secher NJ: Eating disorder that was diagnosed before pregnancy and pregnancy outcome. Am J Obstet Gynecol 2004;190: 206–210.
53 Smith GCS, Shah I, Pell JP, Crossley JA, Dobbie R: Maternal obesity in early pregnancy and risk of spontaneous and elective preterm deliveries: a retrospective cohort study. Am J Pub Health 2007; 97:157–163.
54 Dalziel SR, Parag V, Rodgers A, Harding JE: Cardiovascular risk factors at age 30 following preterm birth. Int J Epidemiol 2007;36:907–915.
55 Ceesay SM, Prentice AM, Cole TJ, Foord F, Weaver LT, Poskitt EM, Whitehead RG: Effects on birth weight and perinatal mortality of maternal dietary supplements in rural Gambia: 5 year randomised controlled trial. BMJ 1997;315:786–790.
56 Kramer MS, Kakuma R: Energy and protein intake in pregnancy. Cochrane Database Syst Rev 2003; CD000032.
57 Haider BA, Bhutta ZA: Multiple-micronutrient supplementation for women during pregnancy. Cochrane Database Syst Rev 2006;CD004905.
58 Makrides M, Duley L, Olsen, SF: Marine oil, and other prostaglandin precursor, supplementation for pregnancy uncomplicated by pre-eclampsia or intrauterine growth restriction. Cochrane Database Syst Rev 2006;CD003402.
59 Hofmeyr GJ, Atallah, AN, Duley L: Calcium supplementation during pregnancy for preventing hypertensive disorders and related problems. Cochrane Database Syst Rev 2006;CD001059.

60 Devi S, Subbulaxmi G, Margetts B, Fall CHD, Potdar RD, Chheda P, Brown N, Sahariah SA, Kehoe S: Processes involved in developing a palatable micronutrient-rich snack for a pre-conceptional and intra-pregnancy micronutrient intervention trial; the Mumbai Maternal Nutrition Project. Early Hum Dev 2007;83(suppl 1):S154.
61 Shankar A, the Multiple Micronutrients Intervention Trial (SUMMIT) Study Group: Effect of maternal multiple micronutrient supplementation on fetal loss and infant death in Indonesia: a double-blind cluster-randomised trial. Lancet 2008;371:215–227.
62 Christian P, Osrin D, Manandhar DS, Khatry SK, de L, Costello AM, West KP: Antenatal micronutrient supplements in Nepal. Lancet 2005;366:711–712.
63 Hawkesworth S, Prentice AM, Fulford AJC, Moore SE: Maternal protein-energy supplementation does not affect adolescent blood pressure in the Gambia. Int J Epidemiol 2009;36:119–127.
64 Belizan JM, Villar J, Bergel E, del Pino A, Di Fulvio S, Galliano SV, Kattan C: Long term effect of calcium supplementation during pregnancy on the blood pressure of offspring: follow up of a randomised controlled trial. BMJ 1997;315:281–285.
65 Vaidya A, Saville N, Shrestha BP, de L Costello AM, Manandhar D, Osrin D: Effects of antenatal multiple micronutrient supplementation on children's weight and size at 2 years of age in Nepal; follow-up of a double blind randomised controlled trial. Lancet 2008;371:492–499.
66 Fall CHD, Stein C, Kumaran K, Cox V, Osmond C, Barker DJP, Hales CN: Size at birth, maternal weight, and non-insulin-dependent diabetes (NIDDM) in South Indian adults. Diabet Med 1998;15:220–227.
67 Yajnik CS, Forrester T, Ramaiya K, Tandon N, Kale S: Prevalence/Incidence, Risk Factors and Future Burden of Type 1, Type 2 and Gestational Diabetes in Developing Countries. Oxford, Oxford University Press, 2008.

Prof. Caroline H.D. Fall
International Paediatric Epidemiology, MRC Epidemiology Resource Centre
University of Southampton, Southampton General Hospital
Tremona Road, Southampton SO16 6YD (UK)
Tel. +44 23 80 777624, Fax +44 23 80 704021, E-Mail chdf@mrc.soton.ac.uk

Newnham JP, Ross MG (eds): Early Life Origins of Human Health and Disease.
Basel, Karger, 2009, pp 42–51

Developmental Origins of Health and Disease: The Importance of Environmental Exposures

Retha Newbold[a] · Jerrold J. Heindel[b]

[a]Developmental Endocrinology and Endocrine Disruptor Section, Laboratory of Molecular Toxicology, Division of Intramural Research, [b]Division of Extramural Research and Training, National Institute of Environmental Health Sciences, Research Triangle Park, N.C., USA

Complex diseases can result from gene-environment interactions. Recent 'epidemics' of chronic diseases like diabetes, childhood asthma, ADHD, and obesity must have an environmental component since genetic modifications cannot account for such large increases over such a relatively short time period. As an example of an environmental component, in utero nutrition is now known to play an important role in susceptibility to cardiovascular diseases and diabetes later in life [1]. This research area paved the way for a new field of study called the 'developmental origins of health and disease' (DOHaD). Initially, DOHaD only focused on nutrition, but animal studies have long documented that the in utero developmental period is a sensitive window for perturbation, not just for nutrition factors, but also for environmental chemicals. Depending on agent, timing, and dose, environmental chemicals can cause death, malformations, low birthweight, or functional changes including increased susceptibility to diseases later in life. Recent reviews [2–4] and entire issues of two journals [5, 6] have been devoted to the topic. These toxicant-induced responses most likely result from altered gene expression or protein regulation associated with abnormal cell proliferation and differentiation involved in interactions between various cell types and the establishment of cell lineages. These changes may lead to abnormal morphological and/or functional characteristics of the tissues, organs, and systems. These alterations may be due, in part, to altered epigenetics and underlining methylation-related protein-DNA relationships associated with chromatin remodeling. Effects may occur in a time-specific (i.e. vulnerable window) and/or tissue-specific manner and the changes may not be reversible. The end result is an animal that is sensitized, rendering it more susceptible to diseases later in life.

Epidemiology data supporting DOHaD combined with data showing alterations in gene expression and tissue imprinting due to in utero exposures to environmental agents provide an attractive framework for integrating the scientific focus on developmental nutrition with developmental chemical exposure. We propose that the DOHaD paradigm be expanded to encompass a broader definition of environment to include exposure to environmental stressors with a focus on environmental chemicals, either alone or in combination with altered nutrition.

The following key points relate to nutrition and chemical exposures during development and serve to support our proposal that the DOHaD paradigm must consider both environmental influences:

- Time-specific (vulnerable window) and tissue-specific effects may occur with both nutritional and environmental chemical exposures.
- The initiating in utero environmental insult (nutritional or environmental chemical) may act alone or in concert with other environmental stressors. That is, in utero exposure could lead by itself to pathophysiology later in life or it could interact with neonatal exposure (same or different environmental stressor(s)) or adult exposure triggering or exacerbating the pathophysiology.
- The pathophysiology may manifest as: the occurrence of a disease that otherwise would not have happened; an increase in risk for a disease that would normally be of lower prevalence; an earlier onset of a disease that would normally have occurred, or an exacerbation of the disease.
- The pathophysiology may have a variable latent period from onset in the prenatal/neonatal period, to early childhood, to puberty, to early adulthood to late adulthood depending on the environmental stressor, time of exposure and tissue/organ affected.
- Altered nutrition and/or exposure to environmental chemicals can lead to aberrant developmental programming that permanently alters tissue, organ or system function. Altered or compromised function (regardless of the stressor – nutritional or chemical) is likely to result from epigenetic changes, e.g. altered gene expression due to effects on imprinting, and the underlying methylation-related protein-DNA relationships associated with chromatin remodeling. The end result is a sensitized individual who is more susceptible to disease later in life.
- The effect of either developmental nutrition or environmental chemical exposures can be transgenerational, affecting future generations.
- While the focus of nutritional changes during development has been on low birthweight, effects of in utero exposure to toxic environmental chemicals or nutritional changes may both occur in the absence of reduced birthweight. The lack of a specific easily measurable biomarker like birthweight makes it more difficult to assess developmental effects. Thus, for both exposures, new more sensitive biomarkers of exposure are needed.
- Extrapolation of risk from both nutritional and chemical exposures may be difficult because effects may not follow a monotonic dose-response relationship. Nutritional effects which result in low birthweight are different from those that

result in high birthweight. Similarly, low dose effects of environmental chemicals may not be the same as the effects that occur at higher doses. Further, environmental chemical and/or nutritional effects may have entirely different manifestations in the embryo, fetus, or perinatal organism, compared to the adult.

- Exposure of one individual to an environmental stressor (environmental chemical or nutritional or combinations) may have little effect, whereas another individual will develop overt disease or dysfunctions due to differences in genetic background including genetic polymorphisms.

DOHaDs nutritional focus originated with human studies and continues to be developed using epidemiology data which are being supported by animal studies. The focus on human studies stems from data on birthweight as an endpoint indicating altered nutrition during development and its association with diabetes, obesity and cardiovascular disease later in life. However, the environmental focus of DOHaD started mainly with animal studies, but there remains a paucity of human data. In general, environmental chemical exposures, especially low-dose exposures, do not necessarily cause altered birthweight, so there is no simple physiological endpoint at birth to indicate environmental exposures. Consequently, animal models are being developed to link environmental exposures to sensitive endpoints such as altered gene expression that lead to altered tissue function resulting in disease later in life. Thus, animal studies are developing the 'biomarkers' of exposure and effect needed for the human studies.

Numerous diseases in animals indicate a role for developmental exposure to environmental agents in their etiology; these include reproductive problems (uterine fibroids, endometriosis, early reproductive senescence, altered fertility and sperm counts), cancers (breast, prostate), and diseases of the cardiovascular, immune, nervous (ADHD, Parkinson's disease) and endocrine (diabetes, obesity) systems [5, 6]. Currently, more diseases in animal models appear to be related to environmental chemical exposures than to altered nutrition.

The focus of our review is on experimental animal studies. Articles cited and diseases discussed are not inclusive but provide examples of reproductive toxicity and obesity to demonstrate 'proof-of-principle'. These studies were chosen to show the state of the science in the area and to highlight data gaps and needs. Our goal is to stimulate interest in integrating basic and epidemiologic data from the two fields, developmental nutrition and developmental toxicology, so that when DOHaD is discussed, the term will encompass all environmental stressors including exposure to environmental chemicals.

Although the focus of this review is not on human studies, it is clear that humans are exposed in utero to a variety of environmental chemical insults, and many are the same chemicals that have been shown to cause increased incidences of disease/dysfunction later in life in animals [7, 8]. Further, adverse effects occur at similar concentrations of exposure in both animals and humans [9]. Thus, the potential exists to extrapolate animal data to human health risk.

Prenatal Exposure to Diethylstilbestrol: Proof of Principle

The classic and prototypical example of a DOHaD phenomenon is developmental diethylstilbestrol (DES) exposure and its long-lasting effects on reproductive tract tissues in both humans and experimental animal models. DES, which was prescribed to pregnant women with high-risk pregnancies in the late 1940s to 1970s, resulted in a rare but significant increase in vaginal neoplasia in their female offspring [10], and numerous other benign reproductive abnormalities in both the male and female offspring. Subsequent epidemiology studies showed that prenatal DES treatment was also associated with increased risk of miscarriage, ectopic pregnancy, infertility, premature birth, uterine fibroids, premature menopause, and increased breast cancer with age [11–13]. Remarkably, every effect noted in DES-exposed humans has been replicated in similarly treated experimental animals. In fact, prenatal DES animal models have also been used to predict effects that were later shown in DES-exposed humans (e.g. early reproductive senescence, paraovarian cysts, oviductal malformations and uterine fibroids in females, and retained testes and lesions in the prostate and seminal vesicles in males).

Although DES is no longer prescribed during pregnancy, it can be used in the laboratory as a prototype estrogenic chemical that acts during development to increase disease incidence later in life. For example, treatment of outbred mice with low doses of DES (0.1 μg/kg) only during neonatal life resulted in altered uterine response and gene expression at puberty; these alterations were irreversible. Further, these neonatally DES-exposed mice developed uterine carcinoma in adulthood. Most importantly, increased susceptibility for uterine tumors was also seen in second-generation DES mice [14]. This finding is of particular importance as the ability to induce a transgenerational effect suggests an epigenetic change occurred during development, and it was transmitted through the germ line. Consistent with this hypothesis, the increased uterine carcinomas observed in DES-treated mice were accompanied by altered methylation of specific estrogen-responsive genes in the uterus [15]. Neonatal DES treatment has also been shown to result in increased and persistent phosphorylation of epidermal growth factor receptor, erbB2 and estrogen receptor-α along with an elevation of c-fos expression, effects capable of activating the receptor-mediated pathways in a ligand-independent manner, thus providing a plausible mechanism for the adverse effects of neonatal DES exposure [16]. It is likely that these effects are related to epigenetic changes in estrogen-responsive genes.

Neonatal treatment of mice with DES was also associated with uterine leiomyomas (fibroids) later in life [17] as previously reported following prenatal treatment. Together, these prenatal and neonatal animal studies provided the background for the recently reported findings in DES-exposed women showing a significant increased risk of prenatal DES being associated with the development of fibroids in adulthood [13]. The etiology of fibroids is currently unclear, but additional investigation into prenatal exposures may provide a link.

Results with other experimental animal models have also shown a link between developmental exposure to DES and uterine fibroids. Early life exposures of the Eker rat to DES increases susceptibility to uterine fibroids because of a defect in a tumor suppressor gene which results in reprogramming of the myometrium leading to an increase in expression of estrogen-responsive genes [18]. Thus, later in life, DES-exposed Eker rats develop uterine fibroids. This experimental animal model provides another example of gene-environment interactions in the DOHaD paradigm.

DES also affects ovarian follicle development in outbred mice if given during the neonatal period; the result is an increase in polyovular follicles with reduced fertilization potential [19]. Neonatal exposure of Sprague-Dawley rats to DES also induces morphological and functional abnormalities in the adult ovary including altered gene expression of steroidogenic enzymes [20].

Thus, DES serves as 'proof of principle' that chemicals with hormonal activity can cause a multitude of diseases/dysfunctions in both animals and humans when exposure occurs during development. The mechanisms involved in DES-induced effects in animals have been shown to be associated with altered gene expression due to altered epigenetic marks. Although similar epigenetic events occur in humans during development, more studies are necessary to show these epigenetic mechanisms are involved in misprogramming for disease later in life.

Other Selected Examples of Adult Diseases with Developmental Origins

Breast Cancer

Breast cancer is the most common cancer in women. Since lifetime exposure to endogenous estrogens is a known risk factor for breast cancer in women, it is plausible that exposure to exogenous estrogens in the form of environmental chemicals will also increase risk of breast cancer. Indeed, there are significant animal data on developmental exposures to endocrine-disrupting chemicals, some of which are estrogenic, and increased cancer risk later in life. Birnbaum and Fenton's review [21] concluded that there is good evidence for increased mammary tumors induced by prenatal or neonatal exposure to a variety of environmental chemicals. Indeed, animal studies have shown that in utero exposures to bisphenol A (BPA; a component of polycarbonate plastic and epoxy resins that line food and beverage containers) [22, 23] which has estrogenic activity, atrazine (a heavily used chlorotriazine herbicide) [21], or dioxin [24] can alter development and organization of the mammary gland in mice and rats; these effects are associated with an increase in susceptibility to mammary cancer later in life [21]. Interestingly, the increased tumor incidence following in utero dioxin exposure was also shown in second-generation pups. Thus, similar to DES [25], in utero exposure to dioxin can have multigenerational effects and may involve epigenetic events.

At present, the exact mechanisms by which alterations in mammary gland development increase cancer risk is unclear. One possibility is that delays in mammary gland development by prenatal exposure to dioxin or atrazine confer an extended window of sensitivity to potential carcinogens after sexual maturity [21]. Additionally, altered morphogenesis may directly lead to neoplastic development. While the mechanisms are uncertain, these studies do suggest that animal and human investigations of breast cancer that focus on adult exposures to environmental chemicals may be trying to correlate exposure and effect at the wrong time; rather, the focus should be on in utero exposures that increase susceptibility to tumor formation later in life. Indeed, Palmer et al. [26] reported an overall 40% increased risk of breast cancer in women exposed in utero to DES, a risk that may increase as these women are just now reaching the age at which the incidence of breast cancer increases.

Testis Function and Fertility

Data relative to the role of exposures to environmental agents in the developmental basis of adult disease come from the work on methoxychlor in male mice. In utero exposure to methoxychlor not only resulted in adverse effects in the adult but these adverse effects were transmitted through the male germ line over at least four generations [27]. The transgenerational effects were only noted when in utero exposure correlated with critical developmental processes of the testes including the time of germ cell remethylation. If dosing occurred later during testis development, the transgenerational effects were absent, but other adverse effects still could be shown in the adult. While other researchers have found significant reproductive effects in adults due to in utero exposures to various environmental chemicals, they either did not look for transgenerational effects, or did not find them perhaps due to dosing too late during development rather than during the time of epigenetic remodeling of germ cell DNA. The transgenerational effect of methoxychlor [27], along with that of dioxin on mammary gland [21] and DES on uterine tumors [25] has the potential to have a major impact on public health and evolutionary biology.

Obesity

Obesity is a disease whose prevalence has risen dramatically in developed countries over the past two to three decades, reaching epidemic proportions in the United States. It is proposed to be caused by prolonged positive energy balance due to a combination of overeating and lack of physical activity on a background of genetic predisposition. The alarming rate of increase in obesity in only two to three decades indicates that the primary cause must lie in environmental and behavioral changes rather than genetics. Indeed, there is an emerging hypothesis that obesity has its origins during development and is influenced by the environment. In this case, environment refers to environmental chemicals now called 'obesogens'. While the role of environmental chemicals in developing obesity may still be an emerging hypothesis, considerable human and animal data link altered nutrition during development to obesity later in life. A large number of epidemiological studies have demonstrated a direct relationship between birthweight and body mass index later in life. These data support the seeming paradox of increased adult adiposity associated with both ends of the birthweight spectrum: higher body mass index with higher birthweight and increased central adiposity with lower birthweight [28]. One mechanism proposed for the effect of altered nutrition during development (either low or high birthweight) on subsequent obesity later in life is alteration of leptin programming [29]. In mice, plasma leptin levels rise transiently during the neonatal period at about 16 days of age. If this leptin surge occurs prematurely via injection of leptin, or due to fetal undernutrition, or a maternal high-fat diet, there is an increase in obesity and leptin resistance in adulthood [30]. It appears that a premature leptin surge alters leptin programming and the developing hypothalamus, which leads to obesity in adulthood. Whether there is a similar leptin surge in human infants is unknown, but if so, there is great potential for translation of animal research in this area to human obesity.

Recent data lend support to the 'obesogen' hypothesis [31]. Baille-Hamilton [32] noted a correlation of the obesity epidemic with increasing exposure to 'man-made' chemicals. Although such data are only correlational, it is tempting to speculate that there is indeed a role for increased exposures to environmental chemicals in the recent epidemic of obesity. It is well-established that many substances including anabolic steroids and DES have been used to promote fattening and growth of animals; further, other chemicals including organophosphate pesticides, carbamates, and antithyroid drugs also cause obesity in animals. Also, there is increasing evidence in animal models that in utero exposure to environmental chemicals at environmentally relevant concentrations alters developmental programming of adipose tissue and/or gastrointestinal-hypothalamic centers. The subsequent obesity observed in these models has been linked to irreversible alterations in tissue-specific function as a result of altered gene expression.

The most likely candidates for altering in utero tissue function in ways that may result in obesity later in life include environmental estrogens such as DES or BPA. Newbold et al. [33] have shown that low doses of DES (1 μg/kg/day), either prenatal or neonatal, caused increased body weight in outbred mice that was not evident at birth but reached significance by 6 weeks of age. At 16 weeks of age, DES-exposed animals had a body fat of 27.6% compared with 20.9% in controls. These DES-treated mice had excessive abdominal fat which has been reported to be associated with cardiovascular disease and diabetes in humans [33]. These mice also had elevated levels of leptin, adiponectin, interleukin-6 and triglycerides that actually developed before the obesity was apparent. Perhaps these endpoints can be used as early biomarkers of subsequent obesity. Increased leptin levels may be due to altered leptin programming due to the environmental chemical exposure. Neonatal exposure to other estrogens, 2OH estradiol, and 4OH estradiol also caused a significantly increased body weight at 4 months of age [33], suggesting that DES is not unique, and that in utero exposures to low doses of environmental agents with estrogenic activity can alter the set point for body weight. In addition, the naturally occurring phytoestrogen genistein (an estrogenic component of soy that acts primarily via ERβ in adipocytes) has also been linked to obesity [34].

In fact, in utero exposure to environmentally relevant doses of BPA, which has estrogenic activity and which has been found in human fetal blood and amniotic fluid at low doses [9], also results in increased body weight of mice [35]. Further, BPA has been shown in vitro to increase glucose transport in preadipocytes [36] and, in combination with insulin, to increase conversion of mouse 3T3-L1 fibroblasts into adipocytes while also increasing lipoprotein lipase activity [37].

Developmental exposure to other environmental chemicals has been linked with obesity. Organotins are persistent and ubiquitous chemicals found as contaminants in fish and shellfish, food crops, wood (antifungal treatment) and stabilizers of polyolefin plastics, all of which can lead to significant human exposures. Organotins are the first of a potentially new class of endocrine-disrupting chemicals that specifically target adipogenesis via the targeting key transcription factors in the adipogenic pathway (retinoid X receptor, peroxisome proliferator-activated receptor-γ). In utero exposure to tributyl tin leads to strikingly elevated lipid accumulation in adipose depots, liver and testis of neonate mice and results in increased epididymal adipose mass in adults [38]; this occurs in the absence of significant weight gain so it is at the expense of muscle mass. Organotins also increase the differentiation of adipocytes in vitro. They act as a potent dual affinity ligand for the retinoid X receptor and the peroxisome activated receptor-γ which play important roles in adipocyte differentiation and energy balance. These transcription factors increase the expression of genes that promote fatty acid storage and decrease expression of genes which induce lipolysis. As a result, they promote insulin sensitivity and increase fat cell mass via increased triglyceride storage. The effects of tributyltin on adipogenesis can be seen in mice at dose levels in the range of human exposures [38].

Thus, there is a growing number of environmental chemicals that have been shown in animal models to increase obesity later in life when administered during development. It is therefore likely that obesity is 'set' based on nutrition and exposures to environmental chemicals during development acting on the genetic background. If this is indeed shown to be true, then the focus on obesity must be changed to prevention by reducing environmental stressors (chemical exposures and nutrition) during development, rather than intervention once obesity has occurred.

Conclusions

Numerous experimental animal models document that developmental exposure to environmental chemicals result in increased susceptibility to diseases later in life.

Further, it is apparent that the developmental basis of adult disease, as stimulated by exposure to environmental chemicals, causes epigenetic changes that lead to altered gene expression and result in altered tissue function; this sets up the tissue for increased susceptibility to disease and dysfunction later in life. Currently, sensitive biomarkers of developmental exposures and early markers of subsequent disease are necessary to connect the data relating environmental exposures to disease in humans. There is also a need for studies to examine the interaction between nutrition and exposures to environmental chemicals in altering susceptibility to diseases later in life.

References

1 Gluckman PD, Hanson MA, Beedle AS: Early life events and their consequences for later disease: a life history and evolutionary perspective. Am J Hum Biol 2007;19:1–19.

2 Miller KP, Gorgeest C, Greenfeld C, et al: In utero effects of chemicals on reproductive tissues in females. Toxicol Appl Pharmacol 2004;198:111–131.

3 Heindel J: Role of exposure to environmental chemicals in the developmental basis of reproductive disease and dysfunction. Semin Reprod Med 2005;24:168–177.

4 Waterland RA, Michels KB: Epigenetic epidemiology of the developmental origins hypothesis. Annu Rev Nutr 2007;27:363–388.

5 Brosen K (ed): Basic and Clinical Pharmacology & Toxicology, Denmark, Wiley-Blackwell Publishing, 2008, vol 102.

6 Knudsen (ed): Reproductive Toxicology, Elsevier Publishing, 2007, vol 23.

7 Califat AM, Xiaoyun Y, Reidy JA, Needham L: Exposure of the U.S. population to Bisphenol A and 4-tertiary-octylphenol: 2003–2004. EHP 2008;116:30–44.

8 Younglai EV, Foster WG, Hughes EG, et al: Levels of environmental contaminants in human follicular fluid, serum and seminal plasma of couples undergoing in vitro fertilization. Arch Environ Contam Toxicol 2002;43:121–126.

9 vom Saal FS, Akingbemi BT, Belcher SM, et al: Chapel Hill bisphenol A expert consensus panel statement: integration of mechanisms, effects in animals and potential to impact human health at current levels of exposure. Reprod Toxicol 2007;24:131–138.

10 Herbst AL, Ulfeder H, Poskanzer DC, Longo LD: Adenocarcinoma of the vagina: association of maternal stilbestrol therapy with tumor appearance in young women. Am J Obstet Gynecol 1999;181:1574–1575.

11 Palmer JR, Hatch EE, Rosenberg CL, et al: Risk of breast cancer in women exposed to diethylstilbestrol in utero: preliminary results (United States). Cancer Causes Control 2002;13:753–758.

12 Swan SH: Intrauterine exposure to diethylstilbestrol: long-term effects in humans. APMIS 2000;108:793–804.

13 Baird DD, Newbold R: Prenatal diethylstilbestrol (DES) exposure is associated with uterine leiomyoma development. Reprod Toxicol 2005;20:81–84.

14 Newbold RR, Bullock BC, McLachlan JA : Uterine adenocarcinoma in mice following developmental treatment with estrogens: a model for hormonal carcinogenesis. Cancer Res 1990;50:7677–7681.

15 Shuanfang L, Hansman R, Newbold R, et al: Neonatal diethylstilbestrol exposure induces persistent elevation of c-fos expression and hypomethylation in its exon-4 in mouse uterus. Mol Carcinog 2003;38:78–84.

16 Miyagawa S, Katsu Y, Watanabe H, Iguchi T: Estrogen-independent activation of ERβ signaling and estrogen receptor α in the mouse vagina exposed neonatally to diethylstilbestrol. Oncogene 2004;23:340–349.

17 Newbold RR, Moore AB, Dixon D: Characterization of uterine leiomyomas in CD-1 mice following developmental exposure to diethylstilbestrol (DES). Toxicol Pathol 2002;30:611–616.

18 Cook J, Davis BJ, et al: Interaction between genetic susceptibility and early-life environmental exposure determines tumor-suppressor-gene penetrance. Proc Natl Acad Sci USA 2005;102:8644–8649.

19 Iguchi T, Kamiya K, Uesugi Y, et al: In vitro fertilization of oocytes from polyovular follicles in mouse ovaries exposed neonatally to diethylstilbestrol. In Vitro 1991;5:359–363.
20 Nagai A, Ikeda Y, Aso T, Eto K, Ikeda MA: Exposure of neonatal rats to diethylstilbestrol affects the expression of genes involved in ovarian differentiation. J Med Dent Sci 2003;50:35–40.
21 Birnbaum LS, Fenton SE: Cancer and developmental exposures to endocrine disruptors. Environ Health Perspect 2003;111:389–394.
22 Vandenberg LN, Maffini MV, Wadia PR, Sonnenschein C, Rubin BS, Soto AM: Exposure to environmentally relevant doses of the xenoestrogen bisphenol A alters development of the fetal mouse mammary gland. Endocrinol 2007;148:116–127.
23 Durando M, Kass L, Viva J, et al: Prenatal bisphenol A exposures induce preneoplastic lesions in the mammary gland in Wistar rats. Environ Health Perspect 2007;115:80–86.
24 Fenton SE, Hamm JT, Birnbaum LS Youngblood GL: Persistent abnormalities in the rat mammary gland following gestational and lactational exposure to 2,3,7,8-tetrachlorodibenzo-p-dioxin (TCDD). Toxicol Sci 2002;67:63–74.
25 Newbold RR, Padilla-Banks E, Jefferson WN: Adverse effects of the model environmental estrogen diethylstilbestrol (DES) are transmitted to subsequent generations. Endocrinology 2006;147:S11–S17.
26 Palmer JR, Hatch EE, Rosenberg CL, et al: Risk of breast cancer in women exposed to diethylstilbestrol in utero: preliminary results (United States). Cancer Causes Control 2002;13:753–758.
27 Hung-Shu C, Anway MD, Rekow SS, Skinner MK: Transgenerational epigenetic imprinting of the male germ-line by endocrine disruptor exposure during gonadal sex determination. Endocrinol 2006;147: 5524–5541.
28 Oken E, Gillman MW: Fetal origins of obesity. Obes Res 2003;11:496–506.
29 Ahima RS: Central actions of adipocyte hormones. Trends Endocrinol Metab 2005;16:307–313.
30 Yura S, Itoh H, Sagawa N, et al: Role of premature leptin surge in obesity resulting from intrauterine undernutrition. Cell Metab 2005;1:371–378.
31 Grun F, Blumberg B: Perturbed nuclear receptor signaling by environmental obesogens as emerging factors in the obesity crisis. Rev Endocr Metab Disord 2007;8:161–171.
32 Baillie-Hamilton PF: Chemical toxins: a hypothesis to explain the global obesity epidemic. J Altern Complement Med 2002;8:185–192.
33 Newbold RR, Padilla-Banks E, Snyder RJ, Jefferson WN: Perinatal exposure to environmental estrogens and the development of obesity. Mol Nutr Food Res 2007;51:912–917.
34 Penza M, Montani C, Romani A, et al: Genistein affects adipose tissue deposition in a dose dependent and gender specific manner. Endocrinology 2006;147:5740–5751.
35 Rubin BR, Murray MK, Damassa DA, King JC, Soto A: Perinatal exposure to low doses of bisphenol A affects body weight, patterns of estrous cyclicity, and plasma LH levels. Environ Health Perspect 2001;109:675–680.
36 Sakurai K, Kawazuma T, Adachi T, et al: Bishenol A affects glucose transport in mouse 3T3 F442A adipocytes. Br J Pharmacol 2004;141:209–214.
37 Masuno H, Kidani T, Sekiya K, et al: Bisphenol A in combination with insulin can accelerate the conversion of 3T3 L1 fibroblasts to adipocytes. J Lipid Res 2002;43:676–684.
38 Grun F, Watanabe H, Samarian Z, et al. Endocrine disrupting organotin compounds are potent inducers of adipogenesis in vertebrates. Mol Endocrinol 2006;20:2141–2155.

Dr. Jerrold J. Heindel
Division of Extramural Research and Training
National Institute of Environmental Health Sciences
Mail Drop EC-23, Research Triangle Park, NC 27709 (USA)
Tel. +1 919 541 0781, Fax +1 919 541 5064, E-Mail heindelj@niehs.nih.gov

Newnham JP, Ross MG (eds): Early Life Origins of Human Health and Disease.
Basel, Karger, 2009, pp 52–64

Developmental Origins of Health and Disease across Generations – Theory, Observation, Experiment

Matthew W. Gillman[a] · George Davey Smith[b] · Mark A. Hanson[c] · Hamish G. Spencer[d]

[a]Obesity Prevention Program, Department of Ambulatory Care and Prevention, Harvard Medical School and Harvard Pilgrim Health Care, Boston, Mass., USA; [b]MRC Centre for Causal Analyses in Translational Epidemiology, Department of Social Medicine, University of Bristol, Bristol, [c]Developmental Origins of Health and Disease Division, Institute of Developmental Sciences, Southampton General Hospital, Southampton, UK; [d]National Research Centre for Growth and Development, Department of Zoology, University of Otago, Dunedin, New Zealand

Most of the chapters in this monograph deal with the effects of an individual's environment in utero on his or her health later in life. Indeed, the previous name for developmental origins of health and disease (DOHaD), fetal origins of adult disease, emphasizes how common chronic diseases can originate during early human development. In this chapter, we extend these concepts to address the extent to which environmental insults during development can have adverse effects not just during the subsequent lifetime of the exposed individual, but on his or her descendants.

This issue is important for public health. Obesity and diabetes are now endemic worldwide and are among the world's leading causes of death and disability. If parents with these conditions transmit them nongenomically to their children and grandchildren, the resulting intergenerational vicious cycle would render efforts to curb diabetes and cardiovascular disease extremely difficult.

The scientific basis of intergenerational induction of disease is being elucidated. Animal studies decades ago demonstrated that experimentally induced gestational diabetes could result in diabetes in the grand-offspring. More recent experimental work has suggested plausible mechanisms for such nongenomic inheritance over generations.

In this chapter, we review this experimental work and its potential relevance for humans. We cite the few human studies that have addressed these issues, and pose research challenges for the future. We put the empirical work in the context both of a theoretical framework as well as importance to public health. Most of the examples in this chapter concern obesity, metabolic disorder, diabetes, and cardiovascular disease, because much of the published work is from those conditions and they are major contributors to human morbidity and mortality.

Definitions

Assume F_0 to be the exposed individual. The exposure could be to a woman during pregnancy, or to a woman or a man before pregnancy. Then F_1 refers to this couple's children, F_2 to their grandchildren, etc. Recall that while the F_1 female is in utero, the gametes that she may contribute to the next generation (F_2) form in her ovary. Therefore a prenatal insult to F_1 female could also directly affect F_2.

Some authors argue that for an exposure to F_0 during pregnancy (affecting F_1 in utero) to have truly transgenerational effects, i.e. effects transmitted via the germ line through multiple generations, one must see them in F_3 [1]. In this chapter, we are not so strict. If insults to the pregnant F_0 affect F_2, we call the effect intergenerational. Likewise, we call it intergenerational if factors influencing nonpregnant F_0 have nongenomic effects on F_1.

Early Animal Experiments Showing Intergenerational Influences

In the early 1970s, Stewart et al. [2] exposed a line of rats to mild protein/energy restriction for 12 generations, including during pregnancies. Not surprisingly, compared with their normally fed counterparts, the rats had lower body weights and lengths and lower organ weights except that the brains were relatively spared. They also exhibited altered behavior and poorer learning. In a companion experiment, however, the investigators restored normal diet after 10 generations either during pregnancy, at birth or at weaning of the offspring [3]. It took 3 generations to realign body weights with the control animals; the first and second generation progeny of the F_0 pregnancy rehabilitation group were actually heavier than the normal controls. After 3 generations, the rats rehabilitated at weaning approached but never quite reached normal body weights. Behavioral outcomes also recovered to a variable degree over several generations. Because both the initial malnourishment and the rehabilitation were not transient but instead were continuous over several generations, these experiments do not unequivocally demonstrate intergenerational effects as we have defined them, but they may be relevant to human migration or socioeconomic transition. In addition, Stewart group's observations emphasize the challenge of reversing developmentally induced phenotypes.

Later in the same decade, also in a rat model, van Assche, Aerts, and colleagues performed classic studies on diabetes during pregnancy. Over several experiments, they induced gestational diabetes in F_0 using streptozotocin, a pancreatic beta-cell toxin [4]. If the gestational diabetes was not severe, they observed the set of consequences hypothesized by Pederson and Freinkel [5]: F_1 fetal hyperglycemia, hyperinsulinemia,

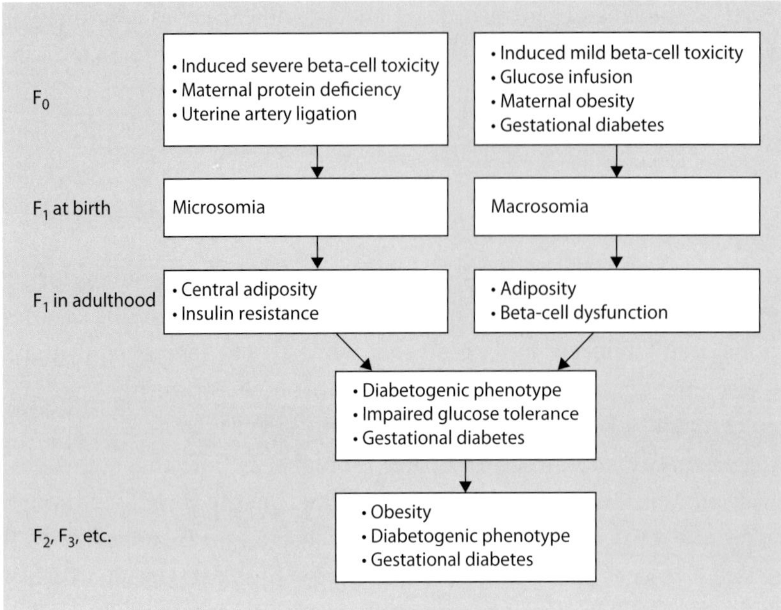

Fig. 1. Schematic representation of conditions producing either micro- or macrosomia resulting in an adult diabetogenic phenotype, which then amplifies across generations through the maternal line.

and macrosomia. The F_1 adult offspring tended to be overweight, hyperphagic, and to have impaired glucose tolerance, which worsened in situations of glucose stress or insulin resistance, such as pregnancy [6]. Glucose infusion into the pregnant F_0 dam or pre-existing F_0 obesity produced a similar F_1 phenotype. In all of these models, the obese, diabetogenic F_1 female tends to get diabetes in her pregnancy, leading to the same adult consequences in F_2, F_3, etc. [4].

Remarkably, a diabetogenic phenotype also emerges among F_1 adults who are microsomic at birth, with insulin resistance as a cardinal feature. Prenatal perturbations producing this set of consequences include severe pharmacologically induced gestational diabetes, protein deficiency, or uterine artery ligation in F_0 pregnant dams. These results are analogous to human observations that both F_0 gestational diabetes (which causes F_1 macrosomia) itself and reduced F_1 fetal growth are associated with type 2 diabetes in F_1. Moreover, they show that regardless of the etiology of the diabetogenic phenotype in F_1, it is transmissible to F_2, F_3, etc. through the maternal line (fig. 1).

The Intergenerational Transmission of Obesity and Diabetes

Irrespective of the mechanisms involved, these observations may have implications for human health. In the developed world, obesity is epidemic and gestational diabetes, for which obesity is the biggest risk factor, is becoming more prevalent [7, 8]. Most, but not all, epidemiologic studies show that gestational diabetes (F_0) is associated with obesity in the offspring (F_1) [9–12]. If the offspring is female, her obesity may lead to gestational diabetes when she becomes pregnant, leading to obesity in F_2, F_3, ad infinitum. In the developing world, not only is obesity becoming epidemic but fetal growth restriction is still common, perhaps leading to a double burden of diabetes in the succeeding generations. According to Wild et al. [13], diabetes prevalence will double in the developing world, primarily India and China, from 2000 to 2030, including among women of reproductive age.

Towards Establishing Effects in Humans

While the potential exists for such intergenerational influences to have major effects on public health for some time to come, the extent to which they actually operate in human populations is not known. Insults applied in animal studies are often orders of magnitude greater than insults that occur in humans. Therefore, one should interpret animal experimental studies as proof of principle and hypothesis generating rather than directly translating them to the human setting.

Prospectively following study samples over multiple generations is a major challenge – it requires multiple generations of investigators! Hints emerge, however, from studies using historical or administrative data. For example, several studies show that F_1 birth weight is associated with F_0 parental cardiovascular morbidity and mortality, with stronger associations for mother than for father [14]. One UK study showed that type 2 diabetes in the maternal (but not paternal) F_0 grandparents was associated with increased F_2 birth size [15]. Another study found that offspring of fathers with type II diabetes had lower birthweight [16].

Multigeneration epidemiologic studies will no doubt continue to emerge, and they will provide a substantial advance in knowledge. Nevertheless, while observational studies have many strengths, they have one major limitation in common, the potential for confounding. Apparent associations of early life factors with health outcomes could actually be due to influences that are socially or economically patterned. For example, many studies show that having been breastfed is associated with lower obesity rates [17]. It is possible that this association is explained by the fact that families who decide to breastfeed typically also exhibit healthful dietary and physical activity practices, which may themselves lead to less obesity [18].

For behaviors in which both mother and father participate, one way of addressing the extent to which the consequences of in utero stimuli explain findings is by comparing the magnitudes of maternal vs. paternal associations. While maternal exposures during pregnancy can have intrauterine impact, paternal exposures during pregnancy cannot. Conversely, both maternal and paternal exposures are likely to be socially patterned in similar ways, and these factors could confound ostensible intergenerational influences. For example, Davey Smith [19] and others have argued that because in some studies maternal and paternal smoking show the same apparent effects on offspring obesity (unlike on birth size), an in utero effect is unlikely. In some circumstances, the comparison of maternal vs. paternal associations may not be an ideal test of common confounding [20], as intergenerational influences due to epigenetic mechanisms such as imprinting (see below) may emanate from the paternal as well as the maternal line [21, 22]. Nevertheless, in situations in which maternal and paternal associations are equivalent, confounding seems a more parsimonious explanation than a paternal-line influence exactly matching a maternal-line influence.

Confounding by genetic inheritance is a persistent DOHaD issue. For example, many studies suggest that maternal (F_0) obesity leads to deleterious consequences for F_1 health outcomes, and we have argued above that these effects may extend to F_2, F_3, etc. But could apparent effects on F_1 (and even subsequent generations) be due to genetic inheritance rather than in utero – or even postnatal – environmental influence? One way to address confounding both by genetic factors and by nongenetic confounding influences takes advantage of the recognition that, during gamete formation, parent to offspring gene transfer is a random process. The resulting observational study design, called Mendelian randomization [23], approximates a randomized controlled trial if an instrumental variable is available as a proxy for the exposure of interest, in this example maternal obesity.

Using the F_0 maternal FTO genotype, controlling for F_1 offspring FTO genotype, as an instrument for maternal obesity, Lawlor et al. [24] did not find that maternal BMI predicted offspring fat mass at age 9–11 years. Despite the relatively low statistical power of that study, its results raise the possibility that inherited genetic influences, not a developmental process, are the major force behind the strong associations between maternal (F_0) and child (F_1) obesity. This interpretation is consistent with the findings of a study in nonpregnant individuals showing that common variation in the FTO gene predicts a wide range of cardiometabolic phenotypes to the extent expected by the association of FTO with BMI [25].

In contrast, in a study using a sib-pair design, prevalence of offspring obesity was lower after maternal bariatric surgery, which produced larger maternal weight loss between pregnancies than before [26]. Also, animal studies generally demonstrate an in utero etiologic role for maternal obesity. For example, using the agouti yellow viable mouse model, Waterland et al. [27] showed that F_0 maternal obesity is amplified through to the F_3 generation. Human sib-pair studies of the associations of (obesity-related) F_0 gestational diabetes with F_1 obesity and type 2 diabetes [9] suggest the same. One possible explanation for the discrepancies among these studies is that the variation in exposure ('dose of intervention') is higher in both the bariatric surgery example and many animal experiments than in the naturally occurring situations, although the study of Waterland et al. [27] was more naturalistic than most animal studies.

One underused approach for estimating unconfounded in utero effects in humans is assessing health outcomes among offspring whose mothers participated in randomized controlled trials during pregnancy [28–32]. The appearance of adequately powered randomized controlled trials in recent years [33–35] makes this strategy attractive for assessing gestational F_0 effects on F_1. Although challenging in terms of logistics and resources, assiduously following F_1 probands until they reproduce will allow assessment of F_0 influences on F_2 outcomes in a more direct manner than previously possible.

Mechanisms of Intergenerational Influences

In the animal experimental literature, many examples now exist of effects of F_0 environmental perturbations during gestation on F_2 outcomes. Perturbations include dietary manipulation, enforced physical activity, exposure to hormones such as glucocorticoids or estrogens or to endocrine disruptors, and developmental disruption following ligation of the uterine artery [36]. Health outcomes have included weight and composition of organs and the body as a whole; markers of metabolic derangement such as elevated blood pressure and endothelial dysfunction [37], insulin resistance, and glucose intolerance; endocrine and reproductive function [38]; development of tumors, and cellular function including enzyme levels and activity.

These perturbations often cause epigenetic changes in DNA methylation or histone configuration [see the chapter by Lillycrop et al., pp. 11–20], which provide potential mechanisms for the observed changes in phenotypic outcomes. Dietary challenges in F_0 produce such epigenetic changes in gene expression which can affect metabolism in F_2 without further challenge in F_1 females [39]. Yet, as discussed above in 'Defintions', for epigenetics to be an unassailable explanation for germ line propagation, the F_3 generation must be affected. Recently, Anway and Skinner [40] showed that administration of a high-dose combination of endocrine disruptors to the F_0 gestating rat dam caused defects in spermatogenesis in a large proportion of progeny from F_1 through F_4. These descendants also had increased tumor development, kidney and prostate disease, and immune abnormalities. The phenotype was passed down through only the male germ line (fig. 2). The high rates of these adverse outcomes (30–90% of animals) rules out standard genetic explanations, e.g. DNA sequence mutations.

Many authors aver that global demethylation occurs after conception and thus epigenetic explanations for intergenerational influences appear illogical. However, a small proportion, perhaps 10%, of transposons (metastable epialleles) do not lose their methylation marks [41]. In addition, it is possible that methylation interacts with other epigenetic processes, perhaps involving imprinting [27]. Moreover, in contrast with other epigenetic changes, imprinted genes appear to be relatively resistant to demethylation [41].

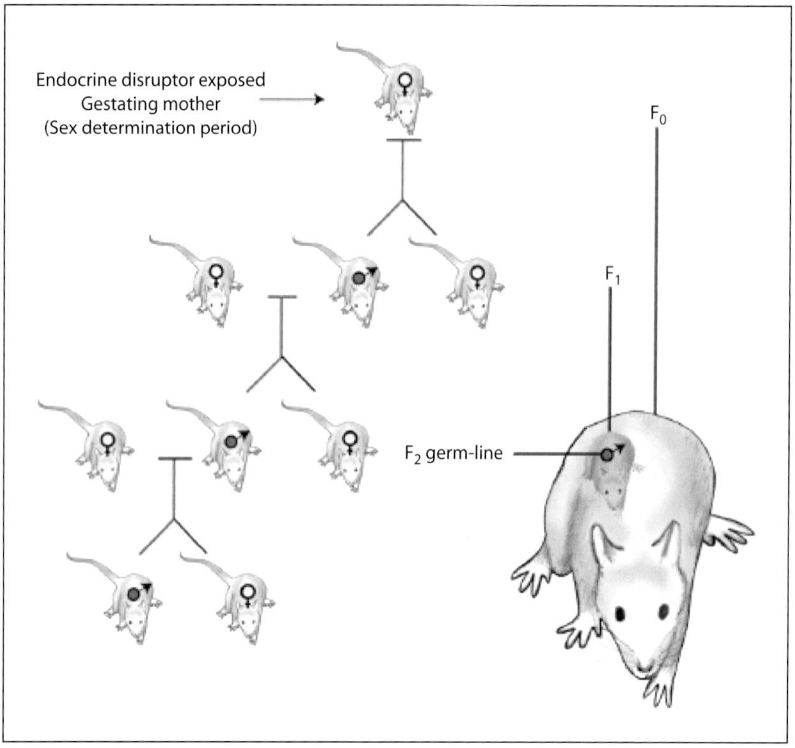

Fig. 2. Schematic of endocrine disruptor-induced transgenerational phenomena via direct exposure of the F_0 mother, F_1 embryo, and F_2 germ line. The F_3 generation is the first without direct exposure. Reproduced with permission from Skinner [1].

The probability that imprinted genes are involved in intergenerational transmission has several implications for DOHaD research. First is that mathematical models predict that there are only several hundred imprinted genes in the human, fewer than the mouse, and thus a potentially tractable number for study [42]. Second, these models predict that the identity of these genes is species-specific, meaning that to investigate human health one has to study human subjects [42]. Third, imprinting follows a different time course in male and female embryos, suggesting that sex differences in major diseases like autoimmune and cardiovascular disease could have very early origins [43].

A fourth implication deals with assisted reproductive technologies, which are becoming more common in developed countries. In 2004, 1% of births in the US and 18% of multiple births were the result of assisted reproduction [44]. One issue is that the in vitro fertilization process exposes oocytes or embryos to factors in culture media that could induce epigenetic changes in both imprinted and nonimprinted genes. The incidence of large offspring syndrome in sheep and rare conditions manifesting abnormal growth in humans, such as Beckwith-Wiedemann syndrome, are consistent with this inference [45]. In addition, the advent of intracytoplasmic sperm injection means that previously infertile men can now procreate. The reason for their relative sterility may be spermatic abnormalities resulting from endocrine disruption, itself a cause of gene imprinting [45]. Thus the success of in vitro fertilization, especially in conjunction with intracytoplasmic sperm injection, in restoring fertility may have long-term deleterious consequences for the children and grandchildren. While no intergenerational studies yet exist, some studies of F_1 offspring of pregnancies conceived through assisted reproductive technologies suggest adverse cardiovascular consequences [46]. These considerations serve as a reminder that it is no longer a 'fact' that the father's entire contribution to his unborn child is half his genome.

Intergenerational influences that arise from epigenetic modifications are not necessarily transmitted through the germ line. Apparently via epigenetic changes in the hippocampus, postnatal F_0 maternal licking and grooming behavior is mimicked by F_1 maternal behaviors [47]. When these F_1 maternal behaviors cause similar epigenetic changes in their pups, F_2 mothers exhibit these behaviors as well. Because the behavior pattern is repeated in each successive generation, these findings provide a biological basis for the well-known phenomenon of intergenerational transmission of behavioral traits.

One underappreciated potential contributor to intergenerational influences is fetal-maternal microchimerism. Not only do the gestating mother and fetus exchange cells regularly, but the exchange is bidirectional and each other's cells persist long term [48]. What are the implications of these observations? When a female F_1 fetus becomes a mother, her F_0 mother's cells could be passed on to her F_2 children, possibly carrying the F_0 mother's environmental (epigenetic?) exposure information with them. It is also possible that the F_0 mother's subsequent children will receive their F_1 older siblings' cells, leading one to speculate on the potential to convey intergenerational influences to F_2 nieces and nephews.

Theoretical Perspectives

A full understanding of intergenerational effects requires us to know something about how and why they have evolved, as well as their evolutionary consequences. Standard evolutionary theory focuses on the long-term effects of natural selection on genes that underlie traits that improve reproductive fitness. Evolutionary biologists have traditionally emphasized individual-level single-generation selection leading to changes in the population-level frequencies of nuclear genes. In contrast, the types of influences we discuss in this chapter involve related individuals and effects across multiple generations. Moreover, the inheritance of such influences is nongenomic, arising from developmental plasticity, which imparts a within-generation phenotypic response to environmental challenges at the level of the individual [49].

Intergenerational transmission of environmental perturbations may have selective advantages in addition to the disadvantages that most of this chapter emphasizes. In the simple world of the bacterium *Bacillus subtilis*, for instance, nutritional limitation induces epigenetic changes that lead to two different types of cells in the second generation, one that sporulates and grows slowly and another that grows faster and does not sporulate [50]. These two subpopulations have different reproductive potentials, depending on the ensuing environment. This response is a form of bet-hedging, ensuring the propagation of the entire clone in an unpredictable environment. It is harder to observe unequivocal advantages in humans. We have two parents, both of whom contribute epigenetic influences but only one of whom gestates. Biological parental-fetal conflict may exist because in theory, each exists to maximize its own reproductive fitness. A great deal of development occurs in utero, so that the fetus must rely on parents to transduce environmental signals. Our lifespans are long, as are those of our descendants.

If environmental effects are passed on unambiguously to generations that have not been exposed, then the implications for our understanding of evolution are profound because such intergenerational effects alter phenotypes, the target of natural selection. As a consequence, the course of evolutionary adaptation may be different from our intuitive forecast, [51] something predicted by even simple mathematical models [52, 53]. Forecasting the implications, however, is difficult. On the one hand, for example, intergenerational effects could hinder classical adaptation of a population to its environment by inducing a time lag into the genetic changes or by masking them from the processes of selection [53]. On the other hand, intergenerational effects may allow organisms to anticipate an environmental challenge and respond appropriately, improving their environmental fit [54].

To model intergenerational influences successfully, theorists must obtain particular sorts of information from experimentalists, whereas other kinds of data may not be relevant. For example, modelers need to know the inheritance patterns of intergenerational effects (do the effects last just one or several generations?), although the exact mechanism, for example methylation or histone modification, may not matter. Exactly what environmental change drives heritable effects is likely to be crucial, as is a view of which parameters experimentalists consider important, such as the proportion of offspring that pass on the effects. Theorists must give something in return. Most fundamentally, they should be able to offer explanations for previously unexplained phenomena and make suggestions for future experiments to clarify properties of the study system.

New phenomena often need new concepts to describe them. In the case of intergenerational influences, some of these concepts parallel standard ideas in genetics, and it will help to have analogous terms. Examples include proportion of offspring exhibiting the affected phenotype, for which the genetic analogy is penetrance; how strongly the affected offspring exhibit the modified phenotype (~expressivity), and the proportion of affected individuals that have affected offspring (~inheritability). Recognizing these concepts allows us to understand their significance and, importantly, to distinguish them from classical genetic phenomena. For instance, the strength of a certain environmental insult may alter the level of 'inheritability' but not that of 'expressivity' or 'penetrance,' and we might want then to investigate what molecular mechanisms underlie such a peculiar phenomenon. Because intergenerational influences have been so relatively neglected, now is an exciting time for researchers interested in such phenomena, and there is great opportunity for productive interactions between theorists and experimentalists.

Conclusions

Numerous animal experiments show that F_0 environmental insults during pregnancy can have phenotypic effects that mimic adverse human health consequences not only on the F_1 offspring but also on their F_2 offspring as well. Some recent experiments suggest that some of these effects are transmitted to multiple subsequent generations through the germ line. The best current explanation for these findings is that they are initiated by F_1 epigenetic processes that, unlike traditional views of most epigenetic changes, resist erasure after F_2 conception.

Adequate data on human populations to address intergenerational influences are understandably scarce because of the long follow-up necessary and, to a certain extent, the limitations of observational studies. At this point, current evidence suggests that diabetes is transmitted this way, although the evidence in favor of such transmission for obesity and cardiovascular disease is weak [20, 55].

It is important to amass data that allow appraisal of intergenerational processes in generating health effects in humans for two major reasons. The first is that epigenetic and other pathophysiologic processes that are evident in animals may or may not exist in humans. The second and more crucial reason is that the 21st century may witness a vicious cycle of intergenerational amplification of obesity, diabetes, and cardiovascular disease. Identifying and quantifying this process will provide clues to preventing it at the source.

References

1 Skinner MK: What is an epigenetic transgenerational phenotype? F3 or F2. Reprod Toxicol 2008;25: 2–6.

2 Stewart RJ, Preece RF, Sheppard HG: Twelve generations of marginal protein deficiency. Br J Nutr 1975;33:233–253.

3 Stewart RJ, Sheppard H, Preece R, Waterlow JC: The effect of rehabilitation at different stages of development of rats marginally malnourished for ten to twelve generations. Br J Nutr 1980;43:403–412.

4 Aerts L, Van Assche FA: Animal evidence for the transgenerational development of diabetes mellitus. Int J Biochem Cell Biol 2006;38:894–903.

5 Freinkel N: Banting lecture 1980. Of pregnancy and progeny. Diabetes 1980;29:1023–1035.

6 Plagemann A, Harder T, Rake A, Voits M, Fink H, Rohde W, Dorner G: Perinatal elevation of hypothalamic insulin, acquired malformation of hypothalamic galaninergic neurons, and syndrome X-like alterations in adulthood of neonatally overfed rats. Brain Res 1999;836:146–155.

7 Dabelea D, Snell-Bergeon JK, Hartsfield CL, Bischoff KJ, Hamman RF, McDuffie RS: Increasing prevalence of gestational diabetes mellitus (GDM) over time and by birth cohort: Kaiser Permanente of Colorado GDM Screening Program. Diabetes Care 2005;28:579–584.
8 Aljohani N, Rempel BM, Ludwig S, Morris M, McQuillen K, Cheang M, Murray R, Shen GX: Gestational diabetes in Manitoba during a twenty-year period. Clin Invest Med 2008;31: E131–E137.
9 Dabelea D, Hanson RL, Lindsay RS, Pettitt DJ, Imperatore G, Gabir MM, Roumain J, Bennett PH, Knowler WC: Intrauterine exposure to diabetes conveys risks for type 2 diabetes and obesity: a study of discordant sibships. Diabetes 2000;49:2208–2211.
10 Hillier TA, Pedula KL, Schmidt MM, Mullen JA, Charles MA, Pettitt DJ: Childhood obesity and metabolic imprinting: the ongoing effects of maternal hyperglycemia. Diabetes Care 2007;30: 287–2292.
11 Gillman MW, Rifas-Shiman SL, Berkey CS, Field AE, Colditz GA: Maternal gestational diabetes, birth weight, and adolescent obesity. Pediatrics 2003;111: e221–e226.
12 Silverman BL, Metzger BE, Cho NH, Loeb CA: Impaired glucose tolerance in adolescent offspring of diabetic mothers. Relationship to fetal hyperinsulinism. Diabetes Care 1995;18:611–617.
13 Wild S, Roglic G, Green A, Sicree R, King H: Global prevalence of diabetes: estimates for the year 2000 and projections for 2030. Diabetes Care 2004;27: 1047–1053.
14 Davey Smith G, Hypponen E, Power C, Lawlor DA: Offspring birth weight and parental mortality: prospective observational study and meta-analysis. Am J Epidemiol 2007;166:160–169.
15 McCarron P, Davey Smith G, Hattersley AT: Type 2 diabetes in grandparents and birth weight in offspring and grandchildren in the ALSPAC study. J Epidemiol Community Health 2004;58:517–522.
16 Hypponen E, Davey Smith G, Power C: Parental diabetes and birth weight of offspring: intergenerational cohort study. BMJ 2003;326:19–20.
17 Owen CG, Martin RM, Whincup P, Davey Smith G, Cook DG: Effect of infant feeding on the risk of obesity across the life course: a quantitative review of published evidence. Pediatrics 2005;115:1367–1377.
18 Owen CG, Martin RM, Whincup PH, Davey Smith G, Gillman MW, Cook DG: The effect of breastfeeding on mean body mass index throughout life: a quantitative review of published and unpublished observational evidence. Am J Clin Nutr 2005;82: 1298–1307.
19 Davey Smith G: Assessing intrauterine influences on offspring health outcomes: can epidemiological studies yield robust findings? Basic Clin Pharmacol Toxicol 2008;102:245–256.
20 Leary SD, Davey Smith G, Rogers IS, Reilly JJ, Wells JC, Ness AR: Smoking during pregnancy and offspring fat and lean mass in childhood. Obesity (Silver Spring) 2006;14:2284–2293.
21 Pembrey ME, Bygren LO, Kaati G, Edvinsson S, Northstone K, Sjostrom M, Golding J: Sex-specific, male-line transgenerational responses in humans. Eur J Hum Genet 2006;14:159–166.
22 Cole TJ, Power C, Moore GE: Intergenerational obesity involves both the father and the mother. Am J Clin Nutr 2008;87:1535–1536.
23 Davey Smith G, Ebrahim S: 'Mendelian randomization': can genetic epidemiology contribute to understanding environmental determinants of disease? Int J Epidemiol 2003;32:1–22.
24 Lawlor DA, Timpson NJ, Harbord RM, Leary S, Ness A, McCarthy MI, Frayling TM, Hattersley AT, Davey Smith G: Exploring the developmental overnutrition hypothesis using parental-offspring associations and FTO as an instrumental variable. PLoS Med 2008;5:e33.
25 Freathy RM, Timpson NJ, Lawlor DA, Pouta A, Ben Shlomo Y, Ruokonen A, Ebrahim S, Shields B, Zeggini E, Weedon MN, Lindgren CM, Lango H, Melzer D, Ferrucci L, Paolisso G, Neville MJ, Karpe F, Palmer CN, Morris AD, Elliott P, Jarvelin MR, Davey Smith G, McCarthy MI, Hattersley AT, Frayling TM: Common variation in the FTO gene alters diabetes-related metabolic traits to the extent expected given its effect on BMI. Diabetes 2008;57: 1419–1426.
26 Kral JG, Biron S, Simard S, Hould FS, Lebel S, Marceau S, Marceau P: Large maternal weight loss from obesity surgery prevents transmission of obesity to children who were followed for 2 to 18 years. Pediatrics 2006;118:e1644–e1649.
27 Waterland RA, Travisano M, Tahiliani KG, Rached MT, Mirza S: Methyl donor supplementation prevents transgenerational amplification of obesity. Int J Obes (Lond) 2008;32:1373–1379.
28 Belizan JM, Villar J, Bergel E, del Pino A, Di Fulvio S, Galliano SV, Kattan C: Long term effect of calcium supplementation during pregnancy on the blood pressure of offspring: follow up of a randomised controlled trial. BMJ 1997;315:281–285.

29 Robinson K, Elliot SJ, Driedger M, Eyles J, O'Loughlin J, Riley B, Cameron R, Harvey D, CHHDP Strategic and Research Advisory Groups: Using linking systems to build capacity and enhance dissemination in heart health promotion: a Canadian multiple-case study. Health Educ Res 2005;20: 499–513.
30 Ben Shlomo Y, Holly J, McCarthy A, Savage P, Davies D, Davey Smith G: Prenatal and postnatal milk supplementation and adult insulin-like growth factor I: long-term follow-up of a randomized controlled trial. Cancer Epidemiol Biomarkers Prev 2005;14:1336–1339.
31 Kinra S, Rameshwar Sarma KV, Ghafoorunissa, Mendu VV, Ravikumar R, Mohan V, Wilkinson IB, Cockcroft JR, Davey SG, Ben Shlomo Y: Effect of integration of supplemental nutrition with public health programmes in pregnancy and early childhood on cardiovascular risk in rural Indian adolescents: long term follow-up of Hyderabad nutrition trial. BMJ 2008;337:a605.
32 Hawkesworth S, Prentice AM, Fulford AJ, Moore SE: Maternal protein-energy supplementation does not affect adolescent blood pressure in The Gambia. Int J Epidemiol 2009;38:119–127.
33 Crowther CA, Hiller JE, Moss JR, McPhee AJ, Jeffries WS, Robinson JS: Effect of treatment of gestational diabetes mellitus on pregnancy outcomes. N Engl J Med 2005;352:2477–2486.
34 Crowther CA, Haslam RR, Hiller JE, Doyle LW, Robinson JS: Neonatal respiratory distress syndrome after repeat exposure to antenatal corticosteroids: a randomised controlled trial. Lancet 2006; 367:1913–1919.
35 Rowan JA, Hague WM, Gao W, Battin MR, Moore MP: Metformin versus insulin for the treatment of gestational diabetes. N Engl J Med 2008;358:2003–2015.
36 Gluckman PD, Hanson MA, Beedle AS: Non-genomic transgenerational inheritance of disease risk. Bioessays 2007;29:145–154.
37 Torrens C, Hanson MA, Gluckman PD, Vickers MH: Maternal undernutrition leads to endothelial dysfunction in adult male rat offspring independent of postnatal diet. Br J Nutr 2009;101: 27–33.
38 Bertram C, Khan O, Ohri S, Phillips DI, Matthews SG, Hanson MA: Transgenerational effects of prenatal nutrient restriction on cardiovascular and hypothalamic-pituitary-adrenal function. J Physiol 2008;586:2217–2229.
39 Burdge GC, Slater-Jefferies J, Torrens C, Phillips ES, Hanson MA, Lillycrop KA: Dietary protein restriction of pregnant rats in the F0 generation induces altered methylation of hepatic gene promoters in the adult male offspring in the F1 and F2 generations. Br J Nutr 2007;97:435–439.
40 Anway MD, Skinner MK: Epigenetic transgenerational actions of endocrine disruptors. Endocrinology 2006;147:s43–s49.
41 Junien C, Nathanielsz P: Report on the IASO Stock Conference 2006: early and lifelong environmental epigenomic programming of metabolic syndrome, obesity and type II diabetes. Obes Rev 2007;8:487–502.
42 Jirtle RL, Skinner MK: Environmental epigenomics and disease susceptibility. Nat Rev Genet 2007;8: 253–262.
43 Vige A, Gallou-Kabani C, Junien C: Sexual dimorphism in non-Mendelian inheritance. Pediatr Res 2008;63:340–347.
44 Wright VC, Chang J, Jeng G, Chen M, Macaluso M: Assisted reproductive technology surveillance – United States, 2004. MMWR Surveill Summ 2007; 56:1–22.
45 Price TM, Murphy SK, Younglai EV: Perspectives: the possible influence of assisted reproductive technologies on transgenerational reproductive effects of environmental endocrine disruptors. Toxicol Sci 2007;96:218–226.
46 Ceelen M, van Weissenbruch MM, Vermeiden JP, van Leeuwen FE, Delemarre-van de Waal HA: Cardiometabolic differences in children born after in vitro fertilization: follow-up study. J Clin Endocrinol Metab 2008;93:1682–1688.
47 Weaver IC, Cervoni N, Champagne FA, D'Alessio AC, Sharma S, Seckl JR, Dymov S, Szyf M, Meaney MJ: Epigenetic programming by maternal behavior. Nat Neurosci 2004;7:847–854.
48 Hall JG: The importance of the fetal origins of adult disease for geneticists. Clin Genet 2007;72: 67–73.
49 Gluckman PD, Hanson MA, Spencer HG, Bateson P: Environmental influences during development and their later consequences for health and disease: implications for the interpretation of empirical studies. Proc Biol Sci 2005;272:671–677.
50 Veening JW, Stewart EJ, Berngruber TW, Taddei F, Kuipers OP, Hamoen LW: Bet-hedging and epigenetic inheritance in bacterial cell development. Proc Natl Acad Sci USA 2008;105:4393–4398.
51 Odling-Smee FJ, Laland KN, Feldman MW: Niche Construction: The Neglected Process in Evolution. Princeton, Princeton University Press, 2003.

52 Gavrilets S: One-locus two-allele models with maternal (parental) selection. Genetics 1998;149: 1147–1152.
53 Spencer HG: Further properties of Gavrilets' one-locus two-allele model of maternal selection. Genetics 2003;164:1689–1692.
54 Gluckman PD, Hanson MA, Spencer HG: Predictive adaptive responses and human evolution. Trends Ecol Evol 2005;20:527–533.
55 Kinra S, Davey Smith G, Okasha M, McCarron P, McEwen J: Is maternal transmission of coronary heart disease risk stronger than paternal transmission? Heart 2003;89:834–838.

Dr. Matthew W. Gillman
Obesity Prevention Program, Department of Ambulatory Care and Prevention
Harvard Medical School and Harvard Pilgrim Health Care
133 Brookline Avenue, 6th floor, Boston, MA 02215 (USA)
Tel. +1 617 509 9968, Fax +1 617 509 9853, E-Mail matthew_gillman@hms.harvard.edu

Newnham JP, Ross MG (eds): Early Life Origins of Human Health and Disease.
Basel, Karger, 2009, pp 65–77

Early Embryo Environment and Developmental Potential

Tom P. Fleming[a] · Kevin D. Sinclair[b] · Karen L. Kind[c]

[a]School of Biological Sciences, University of Southampton, Southampton, [b]School of Biosciences, University of Nottingham, Sutton Bonington, UK; [c]Research Centre for Reproductive Health, School of Agriculture, Food and Wine, University of Adelaide, Adelaide, S.A., Australia

The developing embryo prior to implantation has been shown in recent years to be sensitive to environmental conditions experienced either in vitro or in vivo which may result in changes in phenotype and potential [reviewed by 1–3]. In the context of developmental origins of health and disease, this susceptibility has implications for both fetal and postnatal health and disease risk in later life. Mechanisms by which culture conditions for assisted conception and other reproductive treatments (in vitro) together with maternal physiology and diet (in vivo) may affect embryo developmental potential therefore require close scrutiny. In this chapter, we present three models for the investigation of environmental effects on embryo potential.

Model 1: Oxygen and Embryo Developmental Potential

Oxygen is an important component of the environment of the developing embryo, both in vivo and during ex vivo embryo culture. Lowering the oxygen concentration in the gaseous phase, during embryo culture, from atmospheric levels to levels of 5–7%, improves embryo development, in terms of blastocyst development rate and/or cell number, in a number of species [reviewed by 4]. Increased production of reactive oxygen species has been proposed as the mechanism through which culture at atmospheric oxygen affects embryo development [5, 6]. Oxidative stress may contribute [5], in part, particularly during the pre-compaction stages of development when the embryo appears to be more susceptible to adverse effects of higher oxygen concentration [7]. However, lower oxygen concentrations during embryo culture also more closely resembles in vivo conditions, and may influence embryo development through other mechanisms.

Oxygen tensions of 35–60 mm Hg (5–9%) have been measured in the rabbit, hamster and rhesus monkey oviduct, while uterine levels range from 30–60 mm Hg in hamsters and rabbits to 11–14 mm Hg (1.5–2%) in the rhesus monkey [8]. A higher level of oxygen in the oviduct, compared to the uterus [8], suggests that the embryo encounters a decreasing oxygen gradient as it moves through the reproductive tract, with blastocyst development occurring in the lower oxygen environment of the uterus. The metabolic pathway preference of the embryo shifts from a reliance on oxidative phosphorylation for energy production pre-compaction, to an increased dependence on ATP generation by glycolysis during the post-compaction period. Exposure to a more hypoxic environment may therefore have a physiologically relevant influence on metabolism in the post-compaction stages of embryo development [6]. For the bovine embryo, decreasing the oxygen concentration from 7 to 2% during the compaction to blastulation stages of in vitro production has been reported to increase the development of grade 1 and 2 blastocysts [9], and increase the number of inner cell mass (ICM) cells in the blastocyst [10].

However, the response of the embryo to oxygen concentrations lower than 5% varies between species. For the mouse embryo, in vitro exposure to 2% oxygen represents a hypoxic environment. Mouse embryos cultured in 2% oxygen post-compaction have significantly increased expression of hypoxia-regulated genes, including glucose transporters-1 and -3, when compared to embryos cultured at higher oxygen, or developed in vivo [11]. Following embryo transfer, embryos developed at 2% oxygen have reduced post-implantation viability and those that implant successfully have reduced fetal weight in late gestation [12]. Although placental size is not altered, the surface density of trophoblast available for exchange was reduced following embryo culture at 2%, compared to 7% oxygen, also suggesting effects on placental development [12].

The responses of the mouse embryo to hypoxia illustrate a mechanism through which environmental change can influence embryonic and longer-term development. The embryo detects and responds to changes in its environment through transcriptionally mediated changes in gene expression. Many studies have reported that in vitro embryo development is associated with altered expression of genes involved in key cellular functions, including metabolism, apoptosis, gap junction formation and differentiation [13]. These altered gene expression patterns contribute to adaptive changes in the embryo, such as altered metabolism and cell allocation. Thompson et al. [1] have proposed that exposure of the embryo to an environmental stress influences subsequent development through perturbations in signalling pathways that regulate gene expression patterns. In particular, these causal pathway responses to environmental change may be most important during post-compaction development, when an altered embryonic transcriptome has the ability to influence critical events such as early implantation and placental development [1]. Embryos clearly demonstrate developmental plasticity, in that they are able to adapt to a diverse range of environmental conditions imposed on them during ex vivo manipulations, to maintain cellular and metabolic function, and achieve successful blastocyst development and implantation. Subsequent outcomes, however, may be influenced by the degree of variation in signalling pathways and associated gene and protein [14] expression patterns required to achieve these adaptations. In bovine embryos, pregnancy success is related to the transcriptional profile of the blastocyst [15]. Similarly, while mouse embryos develop to blastocysts under 2% oxygen at similar rates to those cultured at higher oxygen, environmentally induced perturbations in gene expression, cell number and metabolism impact on post-implantation viability, and on placental and fetal development [12].

Differences in the key signalling pathways that mediate gene expression in response to alterations in the extracellular environment may contribute to variation in the response to environmental perturbations between species. Oxygen-regulated gene expression is mediated by transcription factors, the hypoxia-inducible factors (HIFs) [16]. In the bovine embryo, minor differences in gene expression are observed following culture at 2% oxygen, in contrast to 3- to 4-fold changes observed in mouse embryos [10, 11]. HIF-2α is present in the bovine embryo, but the major oxygen-regulated α-subunit HIF-1α has yet to be detected and may be chronically downregulated as an adaptation for development in low oxygen environments [10]. Further analysis of the HIF proteins in the mouse embryo is required to investigate these potential species differences in this signalling pathway.

The significance of hypoxia as a physiological stress during pre-implantation development is unclear. Although embryos may undergo significant variations in environmental oxygen during various ex vivo handling procedures, hypoxic conditions would not normally be encountered. Transfer of pre-compaction embryos into the stimulated uterus, and potentially, the impact of lifestyle factors such as smoking may expose the embryo to an altered in vivo oxygen environment. However, to date, no difference in expression of oxygen-regulated genes was detected in blastocysts exposed in vivo to 6–12 h of maternal hypoxia (10% inspired oxygen) [Kind et al., unpubl.]. Thus, the roles of the HIF transcription factors in mediating gene expression during in vivo embryo development require further study.

In contrast, studies suggest that developing oocytes may be exposed to varying oxygen conditions. The final stages of oocyte growth and maturation occur within antral follicles, with the oocyte separated from a direct vascular supply, and oxygen supplied by diffusion across the follicular cell layers and the fluid-filled antrum. Dissolved oxygen content of 1.5–5% has been reported in human follicular fluid, with variation between women, and between follicles in the same ovary [17]. Other studies report follicular fluid O_2 concentration of 7–17%, with variation between studies potentially related to the differences in collection and analysis methods [18]. Oxygen content and peri-follicular vascularity have both been associated with human oocyte developmental competence. Embryos with the highest implantation potential, and a lower incidence of chromosomal abnormalities, are reported to originate from well-vascularised and oxygenated follicles [17]. In mice, the effect of varying oxygen during oocyte maturation in vitro supports the potential influence of oxygen on oocyte quality [19]. Varying oxygen concentration during murine oocyte maturation, across a range from 2 to 20%, does not alter rates of fertilisation or blastocyst development. However, cell allocation within the blastocyst is altered, with increased total and trophectoderm cells following oocyte maturation at 2%, compared to 20% oxygen. These changes in cell number do not correlate with fetal and placental outcomes, however, as reduced fetal and placental weight was observed following oocyte maturation at 5% oxygen, compared to 20% oxygen or in vivo maturation, respectively [19]. The mechanisms through which varying oxygen exposure during oocyte maturation affects later outcomes require further study. Whether this is related to transcriptionally mediated effects in the maturing cumulus-oocyte complex, or is mediated via epigenetic changes induced by these early environmental perturbations remains to be determined. Other determinants of oocyte competence, including energy production, metabolism and mitochondrial function may also influence subsequent development and require investigation. In vitro oocyte maturation is a developing technology, and is widely used in experimental and livestock reproductive technologies. There is little consensus as to the optimal oxygen concentration for use during in vitro maturation. Studies report benefits of both 5 and 20% oxygen environments, although this varies with species, culture conditions, and the outcomes assessed [reviewed by 4]. Nevertheless, the emergence of in vitro maturation technologies, and the evidence that oocytes encounter varying oxygen conditions in vivo supports further investigation of how oxygen influences oocyte development.

Model 2: Maternal Diet and Embryo Developmental Potential

The effect of maternal diet on early embryo developmental potential has been studied most directly in rodent models and especially the effect of protein undernutrition. By targeting protein undernutrition to the post-mating pre-implantation period exclusively (embryo low-protein diet, Emb-LPD, 9% casein, 3.5 days in mouse; 4.25 days in rats), with control normal diet (NPD, 18% casein) fed for the remainder of gestation and standard chow provided during maternal weaning and throughout offspring life, the consequences for offspring health have been identified [20, 21]. These studies, collectively, have determined that the pre-implantation period is indeed sensitive to maternal protein restriction with postnatal effects identified in increased growth rate to adulthood, hypertension, abnormal anxiety-related behaviour and altered organ allometry. Some of these conditions are gender specific with females generally being more susceptible. Although maternal LPD and global undernutrition models have been applied to other mammalian species during the peri-conceptional period, notably sheep, these studies do not target exclusively the cleavage period. Nevertheless, the consequences of such treatments also result in altered fetal and postnatal growth and cardiovascular dysfunction [22, 23]. Moreover, studies investigating long-term consequences of in vitro culture of pre-implantation embryos as a model for assisted conception and reproductive biotechnologies, using mainly mouse and sheep embryos, further demonstrate adverse effects on fetal or postnatal growth, cardiovascular and metabolic physiology, and behaviour [1, 3, 24–29]. Together, these studies confirm that the pre-implantation period is sensitive to diverse environmental conditions with lasting effects on developmental potential and adult health and disease risk.

The maternal LPD mouse model has been used to determine underlying mechanisms associating protein deprivation with the long-term disease phenotype [21]. Birthweight was increased in response to the Emb-LPD treatment in combined male and female offspring. Whilst excess weight was maintained in females over the next 6 months, this was not the case with males, suggesting gonadal hormone differences in appetite regulation and behaviour between the sexes [30]. Nevertheless, increased perinatal weight was predictive of the later adult weight and susceptibility to disease [21]. These data suggest that the early embryo diet experience leading to increased weight acted as a foundation for later display of abnormality and disease.

To determine the origin of the increase in perinatal weight observed in offspring in response to maternal Emb-LPD, embryo transfer experiments have been conducted whereby blastocysts derived from Emb-LPD- or NPD-fed mothers were placed in opposite horns of recipient NPD-fed dams. Subsequently, at late gestation, it was found that Emb-LPD-derived conceptuses were heavier than NPD-derived ones, indicating that Emb-LPD blastocysts were already 'destined' to an increased growth trajectory irrespective of post-implantation maternal environment [21]. The concept is raised that the pre-implantation interaction with the external maternal tract composition before implantation has a controlling influence on the remaining developmental programme, possibly reflecting a role for nutrient composition in setting of homeostatic metabolic regulators in the embryo that are preserved throughout gestation. This is perhaps not surprising considering the long-term effects that embryo exposure to nutrient levels including amino acids, insulin and cytokines may have on fetal growth [31–33]. Moreover, we have preliminary evidence that Emb-LPD treatment leads to a change in the composition of the uterine fluid [Porter and Fleming, unpubl.].

Our current view is that maternal dietary composition, presumably through plasma intermediates, will lead to a change in the immediate environment experienced by embryos during passage along the maternal tract prior to implantation. This communication is utilised by the blastocyst to activate response mechanisms designed to protect later development from adverse conditions such as those threatened by protein undernutrition. We conjecture that since the blastocyst represents the key stage during which embryonic and extra-embryonic lineages are formed and segregated, this process provides a 'window of opportunity' to modulate the developmental programme and to compensate for environmental deficiencies. Thus, such adaptive responses by the early embryo would concur with a central tenet of the developmental origins of health and disease hypothesis that plasticity in the developmental process may act in a predictive manner to match growth and physiological condition with anticipated postnatal nutrient availability, but that mismatch between predictive and actual postnatal environments underlies the origin of disease [34].

Recently, we have evaluated this concept by investigating extra-embryonic development, responsible for nutrient delivery from mother to conceptus, in the mouse Emb-LPD model. Could the increase in perinatal growth activated by the blastocyst stage in response to maternal protein restriction and known to associate with disease in later life be accountable, at least in part, by plasticity in the extra-embryonic lineages to stimulate nutrient retrieval from the mother? If so, such a response would explain perinatal overweight since a mismatch would exist between a low predictive (pre-implantation) and high actual (post-implantation) nutrient availability. Indeed, when maternal LPD is maintained throughout gestation, perinatal overweight does not result, indicating activation of an extra-embryonic response mechanism where predictive and actual nutrient levels are consistent [21]. However, in such circumstances, while birthweight and postnatal growth rates are normal, onset of adult disease still occurs, indicating that while the embryonic response was beneficial for growth control and perhaps reproductive competitive fitness, it did not guard against later onset of disease [21].

What might be the nature of the putative extra-embryonic response? Following the emergence of the trophectoderm (extra-embryonic) and ICM (mixed progenitor) at the blastocyst stage, the latter subsequently segregates into epiblast (embryonic) and hypoblast (extra-embryonic) lineages. The hypoblast (or primitive endoderm) is the progenitor of the parietal and visceral endoderm layers which in rodents envelop the future embryonic regions of the conceptus. The visceral endoderm develops into the visceral yolk sac which comprises the visceral yolk sac endoderm (VYSE), an absorptive epithelial layer known to endocytose maternal proteins and fluid at its apical surface from the maternal uterine environment, to break these down within active lysosomes, and to deliver released amino acids into the basal, embryonic compartment [35]. These amino acids contribute to protein synthesis during fetal development in parallel with nutrient delivery from the chorio-allantoic placenta formed from trophectoderm derivatives. We have found a major distinction between VYSE behaviour and phenotype dependent upon pre-implantation embryo environment. In conditions of maternal LPD, the VYSE in late gestation, coinciding with the period of maximum fetal growth, engages in an increased rate of endocytosis, is equipped with a more dense population of endocytic cytoplasmic vesicles, and expresses increased levels of megalin protein (the primary multi-ligand receptor for endocytosis [36, 37]) compared with VYSE derived from blastocysts experiencing maternal NPD [21]. Thus, at least one of the extra-embryonic lineages shows strong evidence of plasticity in developmental potential, appearing to activate an adaptive response to early environmental conditions, yet the full mechanisms controlling this change are yet to be identified (fig. 1). Moreover, although the yolk sac forms in a different way within the human, it too is recognised to play a role in nutrient provision during the first trimester [38].

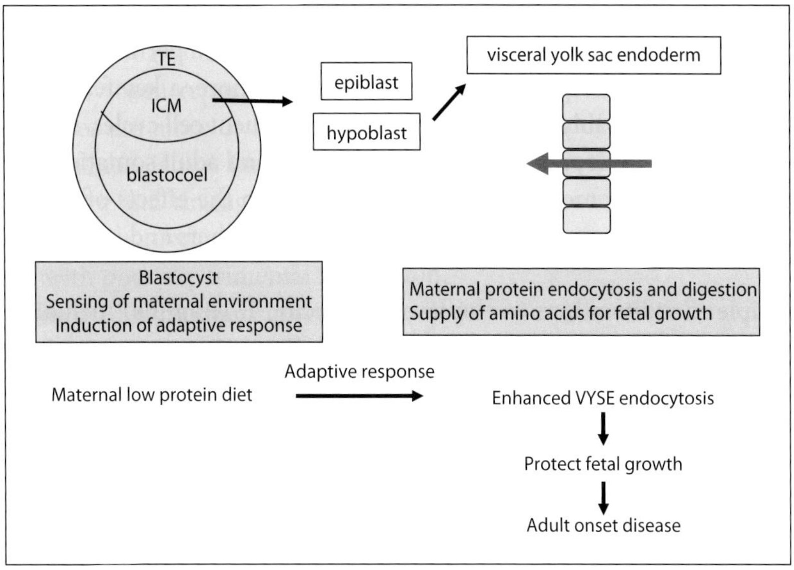

Fig. 1. Schematic showing blastocyst lineages and the proposed adaptive response activated by maternal diet and leading to altered endocytic activity of the VYSE derived from the hypoblast. TE = trophectoderm. From data in reference [21].

Adaptive responses by the embryo to maternal diet likely also occur within the embryonic lineages responsible for generating the fetal tissues. For example, we know that the expression of growth, cardiovascular and metabolic homeostatic regulator genes is affected in fetal or postnatal tissues in response to maternal protein restriction or in vitro culture during pre-implantation development [28, 39, 40]. Further evidence by which early embryo environment alters developmental potential should therefore include analyses of post-implantation morphogenesis and the behaviour and phenotype of cell lineages formed during maturation of the conceptus.

Model 3: B Vitamins, DNA Methylation Programming and Embryo Developmental Potential

Specific components of the maternal diet and of embryo culture media are thought to programme early mammalian development in unknown ways that can predispose adult offspring to a range of diseases [1–3]. There is evidence for heritable changes to DNA methylation at specific loci via nutrients involved in the methyl-donating methionine/folate cycles [41]. Major genome-wide alterations to DNA methylation and post-translational modifications to associated histone proteins are known to occur during oogenesis and pre-implantation development [42], rendering these periods in development sensitive to environmental perturbations [3]. Since

Table 1. Methionine, folate, vitamin B_{12} and B_6 composition of common embryo culture media [53]

Medium	Main species	Methionine nM	Folate nM	Vitamin B_{12} nM	Vitamin B_6 nM	Source
HTF	human	0	0	0	0	Irvine Scientific 90125
G1 medium	human	0	0	0	0	Vitrolife
M16	mouse	0	0	0	0	Sigma M7292
BlastAssist® medium 1	human	0	0	0	0	MediCult
BlastAssist® medium 2	human	50,900	0	0	0	MediCult
ISM1™	human	98,000	0	0	0	MediCult
KSOMaa	mouse/human	50,000	0	0	0	Invitrogen (Gibco)
SOFaa	ruminant	112,500	0	0	0	Sigma B6766
G2 medium	human	50,000	0	0	4,900	Vitrolife
TCM199	ruminant	200,000	2,300	0	0	Sigma M5017
ISM2™	human	103,000	4,500	700	5,900	MediCult
Hams F12	human	30,200	2,950	1,030	291	Invitrogen (Gibco) 21765029
M3	human	24,800	2,490	70	300	MediCult

Table 2. Typical levels of methionine, folate, vitamins B_{12} and B_6 in body fluids of normally fed subjects

Body fluid	Species	Methionine nM	Folate nM	Vitamin B_{12} nM	Vitamin B_6 nM	Reference
Serum	Human	20,000	16	0.3	63[1]	[54]
	Mouse/rat	47,000	17	0.8	750[1]	[55–58]
	Sheep	39,000	7	1.0	626[1]	[43, 59]
Follicular fluid	Human	15,900	18	0.2	36[1]	[54, 60]
	Mouse/rat	23,000	ND	ND	ND	
	Sheep	17,500	14	2	ND	[Sinclair, unpubl.]
Oviductal fluid	Human	50,000	ND	ND	ND	[62]
	Mouse/rat	168,000	ND	ND	ND	[61]
	Sheep	50,000	ND	ND	ND	[63]

ND = To the authors knowledge these values have not been determined.
[1] Pyridoxal phosphate.

References

1 Thompson JG, Mitchell M, Kind KL: Embryo culture and long-term consequences. Reprod Fertil Dev 2007;19:43–52.

2 Sinclair KD, Singh R: Modelling the developmental origins of health and disease in the early embryo. Theriogenology 2007;67:43–53.

3 Fleming TP, Kwong WY, Porter R, Ursell E, Fesenko I, Wilkins A, Miller DJ, Watkins AJ, Eckert JJ: The embryo and its future. Biol Reprod 2004;71:1046–1054.

4 Harvey AJ: The role of oxygen in ruminant preimplantation embryo development and metabolism. Anim Reprod Sci 2007;98:113–128.

5 Guerin P, El Mouatassim S, Menezo Y: Oxidative stress and protection against reactive oxygen species in the pre-implantation embryo and its surroundings. Hum Reprod Update 2001;7:175–189.

6 Harvey AJ, Kind KL, Thompson JG: Redox regulation of early embryo development. Reproduction 2002;123:479–486.

7 Karagenc L, Sertkaya Z, Ciray N, Ulug U, Bahceci M: Impact of oxygen concentration on embryonic development of mouse zygotes. Reprod Biomed Online 2004;9:409–417.

8 Fischer B, Bavister BD: Oxygen tension in the oviduct and uterus of rhesus monkeys, hamsters and rabbits. J Reprod Fertil 1993;99:673–679.

9 Thompson JG, McNaughton C, Gasparrini B, McGowan LT, Tervit HR: Effect of inhibitors and uncouplers of oxidative phosphorylation during compaction and blastulation of bovine embryos cultured in vitro. J Reprod Fertil 2000;118:47–55.

10 Harvey AJ, Kind KL, Pantaleon M, Armstrong DT, Thompson JG: Oxygen-regulated gene expression in bovine blastocysts. Biol Reprod 2004;71:1108–1119.

11 Kind KL, Collett RA, Harvey AJ, Thompson JG: Oxygen-regulated expression of GLUT-1, GLUT-3 and VEGF in the mouse blastocyst. Mol Reprod Dev 2005;70:37–44.

12 Feil D, Lane M, Roberts CT, Kelley RL, Edwards LJ, Thompson JG, Kind KL: Effect of culturing mouse embryos under different oxygen concentrations on subsequent fetal and placental development. J Physiol 2006;572.1:87–96.

13 Lonergan P, Fair T, Corcoran D, Evans ACO: Effect of culture environment on gene expression and developmental characteristics in IVF-derived embryos. Theriogenology 2006;65:137–152.

14 Katz-Jaffe MG, Gardner DK: Embryology in the era of proteomics. Theriogenology 2007;68S:S125–S130.

15 El Sayed A, Hoelker M, Rings F, Salilew D, Jennen D, Tholen E, Sirard MA, Schellander K, Tesfaye D: Large-scale transcriptional analysis of bovine embryo biopsies in relation to pregnancy success after transfer to recipients. Physiol Genomics 2006;28:84–96.

16 Semenza GL: Regulation of mammalian O_2 homeostasis by hypoxia-inducible factor 1. Annu Rev Cell Dev Biol 1999;15:551–578.

17 Van Blerkom J: Epigenetic influences on oocyte developmental competence: perifollicular vascularity and intrafollicular oxygen. J Assist Reprod Genet 1998;15:226–234.

18 Redding GP, Bronlund JE, Hart AL: The effects of IVF aspiration on the temperature, dissolved oxygen levels, and pH of follicular fluid. J Assist Reprod Genet 2006;23:37–40.

19 Banwell KM, Lane M, Russell DL, Kind KL, Thompson JG: Oxygen concentration during mouse oocyte in vitro maturation affects embryo and fetal development. Hum Reprod 2007;22:2768–2775.

20 Kwong WY, Wild AE, Roberts P, Willis AC, Fleming TP: Maternal undernutrition during the preimplantation period of rat development causes blastocyst abnormalities and programming of postnatal hypertension. Development 2000;127:4195–4202.

21 Watkins AJ, Ursell E, Panton R, Papenbrock T, Hollis L, Cunningham C, Wilkins A, Perry VH, Sheth B, Kwong WY, Eckert JJ, Wild AE, Hanson MA, Osmond C, Fleming TP: Adaptive responses by mouse early embryos to maternal diet protect fetal growth but predispose to adult onset disease. Biol Reprod 2008;78:299–306.

22 Edwards LJ, McMillen IC: Periconceptional nutrition programs development of the cardiovascular system in the fetal sheep. Am J Physiol Regul Integr Comp Physiol 2002;283:R669–R679.

23 Gardner DS, Pearce S, Dandrea J, Walker R, Ramsay MM, Stephenson T, Symonds ME: Peri-implantation undernutrition programs blunted angiotensin II evoked baroreflex responses in young adult sheep. Hypertension 2004;43:1290–1296.

24 Thompson JG, Gardner DK, Pugh PA, McMillan WH, Tervit HR: Lamb birth weight is affected by culture system utilized during in vitro pre-elongation development of ovine embryos. Biol Reprod 1995;53:1385–1391.

25 Sinclair KD, McEvoy TG, Maxfield EK, Maltin CA, Young LE, Wilmut I, Broadbent PJ, Robinson JJ: Aberrant fetal growth and development after in vitro culture of sheep zygotes. J Reprod Fertil 1999;116:177–186.

26 Khosla S, Dean W, Brown D, Reik W, Feil R: Culture of preimplantation mouse embryos affects fetal development and the expression of imprinted genes. Biol Reprod 2001;64:918–926.
27 Ecker DJ, Stein P, Xu Z, Williams CJ, Kopf GS, Bilker WB, Abel T, Schultz RM: Long-term effects of culture of preimplantation mouse embryos on behavior. Proc Natl Acad Sci USA 2004;101:1595–1600.
28 Fernandez-Gonzalez R, Moreira P, Bilbao A, Jimenez A, Perez-Crespo M, Ramirez MA, Rodriguez De Fonseca F, Pintado P, Gutierrez-Adan A: Long-term effect of in vitro culture of mouse embryos with serum on mRNA expression of imprinting genes, development, and behavior. Proc Natl Acad Sci USA 2004;101:5880–5885.
29 Watkins AJ, Platt D, Papenbrock T, Wilkins A, Eckert JJ, Kwong WY, Osmond C, Hanson MA, Fleming TP: Mouse embryo culture induces changes in postnatal phenotype including raised systolic blood pressure. Proc Natl Acad Sci USA 2007;104:5449–5454.
30 Toufexis D: Region- and sex-specific modulation of anxiety behaviours in the rat. J Neuroendocrinol 2007;19:461–473.
31 Lane M, Gardner DK: Differential regulation of mouse embryo development and viability by amino acids. J Reprod Fertil 1997;109:153–164.
32 Kaye PL, Gardner HG: Preimplantation access to maternal insulin and albumin increases fetal growth rate in mice. Hum Reprod 1999;14:3052–3059.
33 Sjoblom C, Roberts CT, Wikland M, Robertson SA: Granulocyte-macrophage colony-stimulating factor alleviates adverse consequences of embryo culture on fetal growth trajectory and placental morphogenesis. Endocrinology 2005;146:2142–2153.
34 Gluckman PD, Hanson MA: Living with the past: evolution, development, and patterns of disease. Science 2004;305:1733–1736.
35 Beckman DA, Lloyd JB, Brent RL: Quantitative studies on the mechanisms of amino acid supply to rat embryos during organogenesis. Reprod Toxicol 1998;12:197–200.
36 Christensen EI, Verroust PJ: Megalin and cubilin, role in proximal tubule function and during development. Pediatr Nephrol 2002;17:993–999.
37 Fisher CE, Howie SE: The role of megalin (LRP-2/Gp330) during development. Dev Biol 2006;296: 279–297.
38 Burton GJ, Hempstock J, Jauniaux E: Nutrition of the human fetus during the first trimester – a review. Placenta 2001; 22(suppl A):S70–S77.
39 Kwong WY, Miller DJ, Ursell E, Wild AE, Wilkins AP, Osmond C, Anthony FW, Fleming TP: Imprinted gene expression in the rat embryo-fetal axis is altered in response to periconceptional maternal low protein diet. Reproduction 2006;132:265–277.
40 Kwong WY, Miller DJ, Wilkins AP, Dear MS, Wright JN, Osmond C, Zhang J, Fleming TP: Maternal low protein diet restricted to the preimplantation period induces a gender-specific change in hepatic gene expression in rat fetuses. Mol Reprod Dev 2007;74: 48–56.
41 Waterland RA, Jirtle RL: Transposable elements: targets for early nutritional effects on epigenetic gene regulation. Mol Cell Biol 2003;23:5293–300.
42 Trasler JM: Gamete imprinting: setting epigenetic patterns for the next generation. Reprod Fertil Dev 2006;18:63–69.
43 Sinclair KD, Allegrucci C, Singh R, et al: DNA methylation, insulin resistance, and blood pressure in offspring determined by maternal periconceptional B vitamin and methionine status. Proc Natl Acad Sci USA 2007;104:19351–19356.
44 Midttun O, Hustad S, Schneede J, Vollset SE, Ueland PM: Plasma vitamin B-6 forms and their relation to transsulfuration metabolites in a large, population-based study. Am J Clin Nutr 2007;86:131–138.
45 Pitkin RM: Folate and neural tube defects. Am J Clin Nutr 2007;85:285S–288S.
46 McMillen IC, Robinson JS: Developmental origins of the metabolic syndrome: prediction, plasticity, and programming. Physiol Rev 2005;85:571–633.
47 Pogribny IP, Ross SA, Wise C, et al: Irreversible global DNA hypomethylation as a key step in hepatocarcinogenesis induced by dietary methyl deficiency. Mutat Res 2006;593:80–87.
48 Reed MC, Nijhout HF, Sparks R, Ulrich CM: A mathematical model of the methionine cycle. J Theor Biol 2004;226:33–43.
49 Nijhout HF, Reed MC, Budu P, Ulrich CM: A mathematical model of the folate cycle: new insights into folate homeostasis. J Biol Chem 2004;279:55008–55016.
50 Smith AD: Folic acid fortification: the good, the bad, and the puzzle of vitamin B-12. Am J Clin Nutr 2007;85:3–5.
51 Yajnik CS, Deshpande SS, Jackson AA, et al: Vitamin B_{12} and folate concentrations during pregnancy and insulin resistance in the offspring: the Pune Maternal Nutrition Study. Diabetologia 2008;51:29–38.
52 Selhub J, Morris MS, Jacques PF: In vitamin B12 deficiency, higher serum folate is associated with increased total homocysteine and methylmalonic acid concentrations. Proc Natl Acad Sci USA 2007;104: 19995–20000.
53 Steele W, Allegrucci C, Singh R, et al: Human embryonic stem cell methyl cycle enzyme expression: modelling epigenetic programming in assisted reproduction? Reprod Biomed Online 2005;10:755–766.

54 Steegers-Theunissen RP, Steegers EA, Thomas CM, et al: Study on the presence of homocysteine in ovarian follicular fluid. Fertil Steril 1993;60:1006–1010.
55 Wei IL: The influence of dietary restriction on vitamin B-6 vitamer distribution and on vitamin B-6 metabolizing enzymes in rats. J Am Coll Nutr 1999; 18:144–151.
56 Mackey AD, Lieu SO, Carman C, Gregory JF III: Hydrolytic activity toward pyridoxine-5′-beta-D-glucoside in rat intestinal mucosa is not increased by vitamin B-6 deficiency: effect of basal diet composition and pyridoxine intake. J Nutr 2003;133:1362–1367.
57 Maloney CA, Hay SM, Rees WD: Folate deficiency during pregnancy impacts on methyl metabolism without affecting global DNA methylation in the rat fetus. Br J Nutr 2007;97:1090–1098.
58 Adachi K, Izumi M, Osano Y, et al: Polyamine concentrations in the brain of vitamin B12-deficient rats. Exp Biol Med (Maywood) 2003;228:1069–1071.
59 Coburn SP, Mahuren JD, Guilarte TR: Vitamin B-6 content of plasma of domestic animals determined by HPLC, enzymatic and radiometric microbiological methods. J Nutr 1984;114:2269–2273.
60 Boxmeer JC, Brouns RM, Lindemans J, et al: Preconception folic acid treatment affects the microenvironment of the maturing oocyte in humans. Fertil Steril 2008;89:1766–1770.
61 Harris SE, Gopichandran N, Picton HM, Leese HJ, Orsi NM: Nutrient concentrations in murine follicular fluid and the female reproductive tract. Theriogenology 2005;64:992–1006.
62 Casslen BG: Free amino acids in human uterine fluid. Possible role of high taurine concentration. J Reprod Med 1987;32:181–184.
63 Walker SK, Hill JL, Kleemann DO, Nancarrow CD: Development of ovine embryos in synthetic oviductal fluid containing amino acids at oviductal fluid concentrations. Biol Reprod 1996;55:703–708.

Prof. Tom P. Fleming
School of Biological Sciences, University of Southampton
Bassett Crescent East, Southampton SO16 7PX (UK)
Tel. +44 2380 594398, Fax +44 2380 594459
E-Mail tpf@soton.ac.uk

Newnham JP, Ross MG (eds): Early Life Origins of Human Health and Disease.
Basel, Karger, 2009, pp 78–88

Early Environmental Influences on Lung Development: Implications for Lung Function and Respiratory Health throughout Life

Richard Harding[a] · Ken Snibson[b] · Megan O'Reilly[a] · Gert Maritz[c]

[a]Department of Anatomy and Developmental Biology, Monash University, Clayton, Vic., [b]School of Veterinary Science, University of Melbourne, Melbourne, Vic., Australia; [c]Department of Medical Biosciences, University of the Western Cape, Bellville, South Africa

In humans, the lung develops during prenatal life and infancy, after which it increases in size but not complexity. Exposure to sub-optimal environmental conditions during these early life stages can alter lung development, leading to reduced lung function and an increased risk of respiratory illness later in life. Environmental factors prevailing during early life that have been linked to long-term changes in lung structure, lung function and respiratory health include undernutrition, preterm birth, reduced intrathoracic space, respiratory infections, maternal tobacco smoking and exposure to allergens. In this chapter, we briefly review the impact of some of the major factors.

Fetal Growth and Lung Development

Intrauterine growth restriction (IUGR) affects 8–10% of all births. A major cause is reduced oxygen or nutrient supply to the fetus resulting from maternal vascular diseases, placental pathology, maternal undernutrition or drug use. IUGR is an established risk factor for respiratory complications in both term and preterm neonates [1]. These neonatal problems are not likely to be a result of surfactant deficiency [2, 3], but are more likely a result of delayed clearance of lung liquid or structural immaturity of lung tissue. IUGR has also been associated with reduced lung function in infants [4], children [5], and adults [6], indicating that it can program the lung for altered function throughout life.

Experimental studies have shown that IUGR alters the development of lung parenchyma and airways, and that these effects can persist to maturity. Such studies show that the degree and type of nutrient restriction, as well as its gestational timing, affect the final structure of the lungs. In sheep, the effects of IUGR on lung structure have been studied at 3 stages of life: near-term fetuses, 8-week-old lambs and young adults

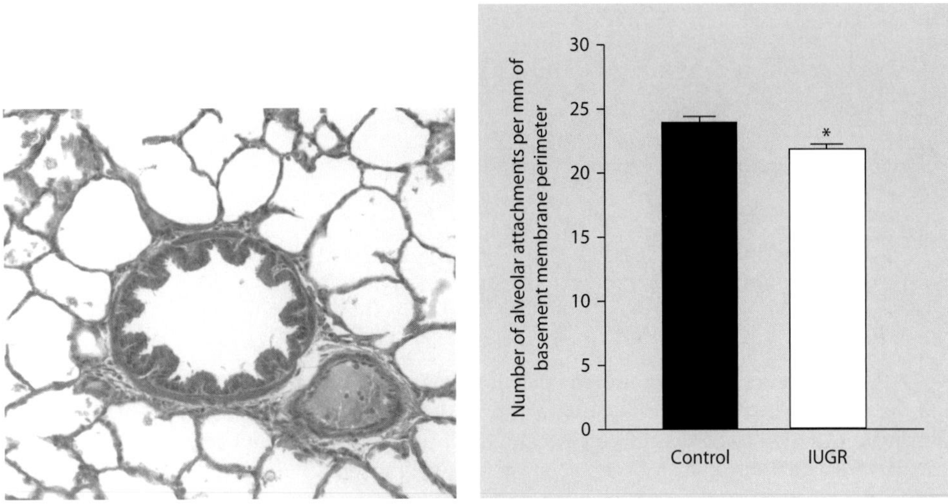

a b

Fig. 1. Effects of IUGR on the number of alveolar attachments on bronchioles of adult sheep. **a** A bronchiole and alveolar attachments in the lungs of an adult sheep. **b** Number of attachments per mm of basement membrane perimeter in adult IUGR (n = 12) and control (n = 8) sheep. The number of attachments was approximately 10% lower in adult IUGR sheep than in control sheep (* $p < 0.05$).

[3, 7–9]. At 8 weeks after birth, IUGR animals had fewer, larger alveoli, with thicker septa and blood-gas barriers [9], resulting in impaired lung function [3]. The reduced number of alveoli and thicker blood-gas barriers were still evident in adult IUGR sheep [8]; fenestrations in the alveolar walls of the adult IUGR sheep suggest accelerated aging of the lungs. These studies show that adequate nutrition during early life is necessary for normal alveolar development and that the deleterious effects of impaired fetal nutrition can persist into adulthood.

IUGR can also affect the conducting airways. In fetal sheep, IUGR led to thinner airway walls and fewer submucosal glands [7]. In adult sheep exposed to IUGR as fetuses, the structure of the airway walls was not different to that in adult control sheep; however, the number of alveolar attachments to bronchioles (per mm of basement membrane) in airways whose circumference ranged from 500–2,000 μm was significantly reduced by about 10% (fig. 1). This reduction in bronchiolar tethering in adults exposed to IUGR, likely a result of a reduced number of alveoli [10], could contribute to the reduced lung function (i.e. reduced FEV_1) reported in adults born with a low birthweight [6].

There are likely to be a number of factors underlying the effects of IUGR on lung development, including fetal hypoxia, reduced nutrient availability and endocrine factors such as elevated plasma levels of glucocorticoids. Fetal lung development is also affected by the physical environment, notably the degree of expansion of the fetal lung and fetal breathing movements [11]. IUGR is associated with diminished fetal breathing, which could contribute to impaired fetal lung development.

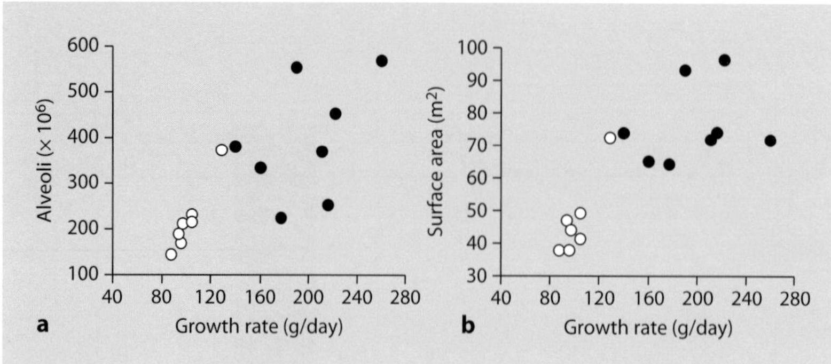

Fig. 2. Significant correlations between postnatal growth rate and the adult number of alveoli (**a**) and the alveolar surface area (**b**) in adult sheep. In these animals, differences in postnatal growth rates were assumed to be due to nutritional differences after birth. Open and filled circles show, respectively, data from slower growing and faster growing sheep.

Postnatal Nutrition, Growth and Lung Development

After birth, the lung continues to mature both structurally and functionally, with the development of new alveoli continuing for 1.5–3 years after birth. Nutrition during this early postnatal period is likely to influence lung development. The importance of early nutrition is supported by the finding that size at 1 year is a significant predictor of later deaths from respiratory cause [12]. Lung development in preterm infants may be especially vulnerable to impaired nutrition, which can arise as a result of limited fat deposits, parenteral feeding, poor suckling ability or gut immaturity. As the lungs are still at an early stage of development in preterm infants, undernutrition or a lack of necessary micronutrients, could have detrimental long-term effects on lung antioxidant and defence mechanisms and alveolar formation. Undernutrition may also affect the innate immunity of the lung. Human and animal studies have shown that malnutrition causes impaired macrophage function, mucociliary clearance, and specific B and T cell responses to infection [13]. Such effects could underlie the increased incidence of respiratory infections in undernourished infants and children.

Animal studies have shown that early postnatal undernutrition can exert persistent effects of alveolarization. In rats, intermittent postnatal starvation coinciding with saccular and alveolar phases of lung development resulted in enlarged alveoli, thicker septa and reduced elastin deposition [14]. Even in long gestation species such as primates and sheep, it is likely that postnatal nutrition can have persistent effects on lung development; this is because the distal regions of the lung continue to develop after birth. A recent study in mature sheep has shown that slow postnatal growth can result in reduced numbers of alveoli and a reduced surface area for gas exchange in relation to lung or body weight [15]. In this study, postnatal growth rate was significantly correlated with the adult number of alveoli and alveolar surface area (fig. 2).

The conducting airways may also be affected by undernutrition. Children who were undernourished in infancy and early childhood are reported to show impaired lung function [16]. A recent study in adult sheep showed that airway wall structure was altered in sheep that grew more slowly than normal sheep after birth [17]. Such changes could contribute to the long-term programming effects of early postnatal undernutrition on lung function.

The Effects of Maternal Smoking on Lung Development

Numerous studies have shown that maternal smoking during pregnancy results in a persistent deficit in lung function of children, most commonly due to reduced flow in the small airways [18]. It is likely that nicotine, which readily crosses the placenta, plays a major role in causing these effects [10]. Altered alveolarization and abnormalities in airway development have been found after gestational exposure to nicotine [19, 20].

Alveolarization

Exposure to tobacco smoke and nicotine are thought to adversely alter lung parenchymal development largely via effects on pulmonary fibroblasts. Fibroblasts play a critical role in alveolarization, during which there is substantial proliferation of interstitial fibroblasts. The formation of elastin by fibroblasts is thought to be critically involved in alveolarization by providing structural support for new secondary septa. Cigarette smoke inhibits fibroblast proliferation and migration by increasing cell cycle transit time; consequently alveolarization is reduced. Smoke exposure also compromises fibroblast-induced repair responses which may contribute to lung disease [21]. In vivo studies also show that nicotine exposure during development can permanently suppress energy metabolism in the lung [22]. It is therefore plausible that nicotine adversely affects the long-term maintenance of lung structure. The type I alveolar epithelial cell (AEC), for example, depends on glycolysis for the supply of ATP required for the membrane-linked Na^+-K^+ ATPase pump [23] that plays a vital role in maintaining cell volume; reducing its activity by inhibition of glycolysis results in cell swelling and the formation of membrane blebs [24]. Inhibition of glycolysis by nicotine will therefore interfere with the ability of type I AECs to maintain cell volume; this is characterized by cell membrane blebbing and rupture (fig. 3). Type II AECs are more numerous in the lungs of nicotine-exposed animals, which is thought to be a response to the loss of type I AECs [25].

It appears that the negative impact of maternal nicotine exposure during gestation and lactation on the growth, development and repair processes of the lungs of offspring is such that lung structure ages more rapidly than in nonexposed animals. One possible reason is the permanent decrease in the glycolytic activity in the lungs

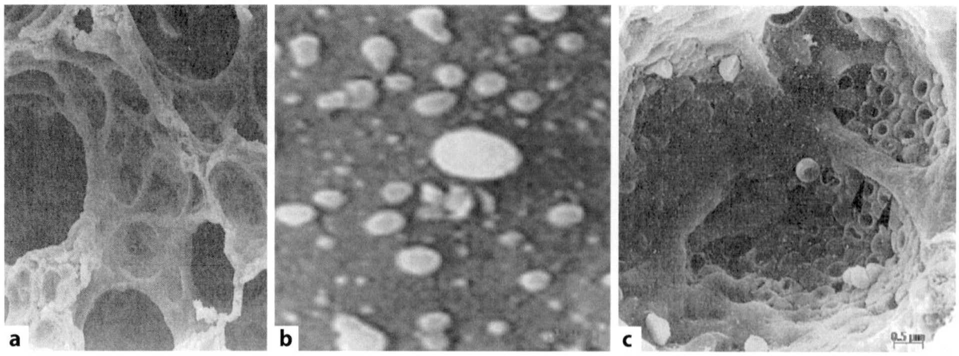

Fig. 3. Scanning electron micrographs of the alveolar surface showing control lung (**a**), blebbing of the alveolar type I cell membrane following nicotine exposure (**b**) and rupture of the alveolar surface, revealing the underlying blood capillary (**c**).

induced by nicotine exposure [26]. Recent studies indeed show that lungs of rat pups exposed to nicotine develop microscopic emphysema after nicotine withdrawal (fig. 4). The elastic tissue framework of the lungs of these animals is also compromised [27]. Exposure to nicotine during the late saccular/early alveolar phase of lung development results in an increase in the size and volume density of the primitive alveoli, decreasing the alveolar surface area for gas exchange [28]. The structural changes in the lungs of these animals resemble faster aging and may make the lungs more susceptible to disease in later life. This is strong evidence of programming of lung structure and function by nicotine exposure during critical stages of lung development [29].

Airway Development

Evidence from epidemiologic studies indicates that airway function of infants whose mothers smoked during pregnancy had a reduction of ~10% in expiratory flow parameters [30]. Maternal smoking is one of the risk factors for the development of airway hyper-responsiveness in children; such a link is supported by animal studies showing altered airway structure and function [19, 31]. Studies in mice suggest that in utero cigarette smoke exposure affects airway reactivity by modulating the lung cyclic AMP levels through changes in phosphodiesterase-4D activity. Importantly, these effects are independent of significant mucous production or leukocyte recruitment into the lung [32].

Preterm Birth, Lung Development and Later Lung Function

Preterm birth (birth before 37 weeks of gestation) occurs in 8–10% of all births and is a major cause of low birthweight. With medical advances, infants born at only

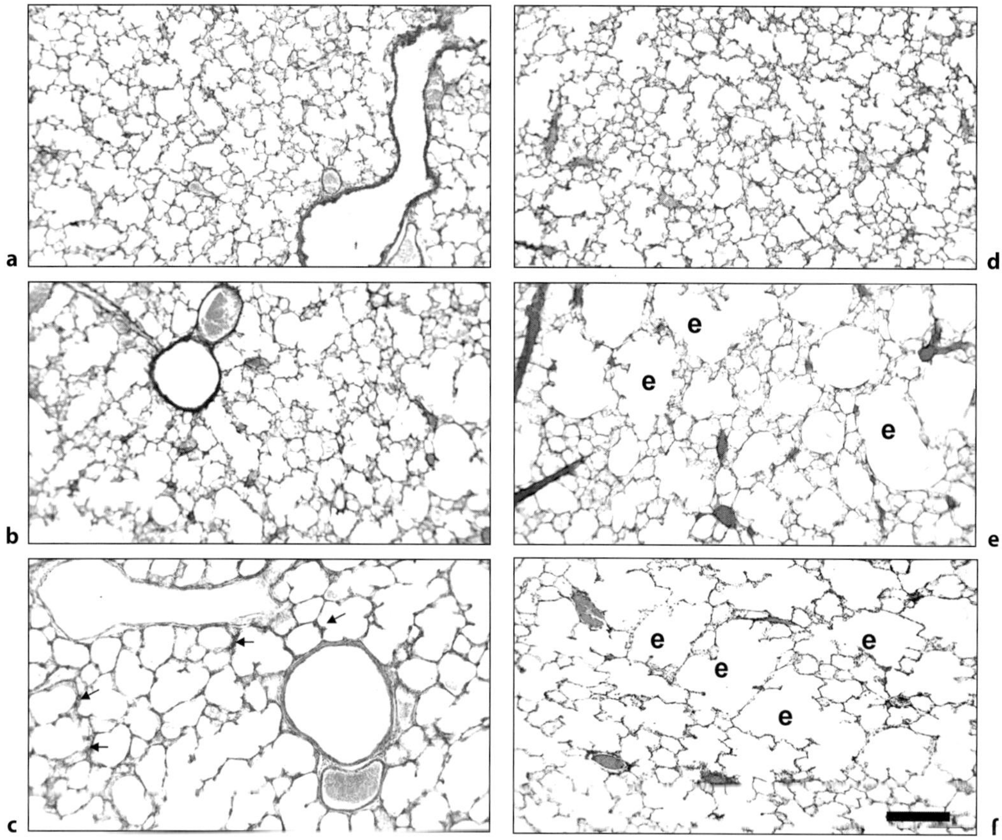

Fig. 4. The separate influences of maternal nicotine exposure either during gestation and lactation (GL), or from the onset of alveolar formation on postnatal day 4 (PNE), on the parenchyma of the lungs of the offspring. **a–c** Light micrographs of the lung parenchyma of 14-day-old control (**a**), PNE (**b**) and GL (**c**) rats. **d–f** Light micrographs of lung parenchyma of 42-day-old control (**d**), PNE (**e**) and GL (**f**) rats. The alveolar walls of the 14-day-old GL rats were thicker (arrows) than those of the control and PNE animals. At postnatal day 42, the lung parenchyma of the G & L and PNE animals showed signs of emphysema (e). Bar = 20 µm.

22–23 weeks of gestation are now capable of survival. The lung can be permanently altered by preterm birth, especially very preterm birth (<30 weeks), and respiratory impairments are common in survivors during childhood, adolescence and probably into adulthood [33]. In severe prematurity, prolonged ventilatory support has been associated with hyaline membrane disease and bronchopulmonary dysplasia (BPD) in infancy, which have been linked to chronic respiratory illnesses such as asthma and chronic obstructive pulmonary disease [34]. Even mild-moderate preterm birth has been associated with adverse respiratory outcomes. For example, healthy preterm infants (mean gestational age, 29.5 weeks) showed reduced functional residual capacity and specific lung compliance, impaired gas mixing efficiency and higher dead space ventilation compared with infants born at term [35]. This suggests that even mild preterm birth, not requiring ventilatory support, can affect alveolarization and elastic fibre deposition. Similarly, mild preterm birth caused an increased incidence of wheeze and reduced expiratory flows in children born at ~35 weeks, suggesting that airway function had been altered [5].

Lung Pathology following Preterm Birth

Before the widespread use of exogenous surfactant, the lungs of survivors of very preterm birth and BPD contained fewer alveoli, a decrease in lung surface area and an increase in bronchiolar smooth muscle and glands [36]. Such infants showed lung hyperinflation, emphysema, and interstitial densities; the principal pathological findings at autopsy are dilated, simplified distal air sacs [36]. Preterm infants who develop BPD and chronic lung disease today are usually those born at earlier gestational ages than in the past (e.g. 23–26 weeks), with extremely low birthweight (<1,000 g); these infants can experience airway remodelling and chronic lung disease [37]. Animal studies have shown that prolonged ventilation of the preterm lung induces simplified distal air sacs due to altered elastogenesis and alveolar secondary septation [38, 39]. A consequence of failed septation is a reduced surface area for gas exchange, particularly the capillary surface area [40].

As discussed above, even mild preterm birth can affect lung development. Using an ovine model of mild-moderate preterm birth, it was found that simply being born preterm was associated with alterations in lung structure, some of which persisted into postnatal life [41, 42]. The preterm lambs had reduced tropoelastin expression in the lungs, and thicker alveolar walls, airway epithelium and blood-air barriers compared to lambs born at term [42]. At 8 weeks, the airway epithelium remained thicker due to cellular hyperplasia and hypertrophy [42]. This mild-moderate preterm birth also led to a persistent delay in the normal postnatal changes in the proportions of type I and type II AECs [41]; in particular, the proportion of type II AECs did not increase after birth as much as in lambs born at term, and this was associated with a transient decrease in surfactant protein expression.

It is now recognized that many preterm infants have been exposed to placental insufficiency and IUGR [43]; that is, their low birthweight is a result of both preterm birth and IUGR. The lungs of such individuals may therefore be doubly affected; for example, preterm birth in IUGR lambs resulted in a reduced diffusion capacity and thus impaired pulmonary gas exchange, compared to term-born IUGR animals [44]. A common mechanism for altered lung development as a result of both IUGR and preterm birth could be impaired nutrition during the period of alveolarization. Hypoxia may also play a role, as both IUGR and preterm birth are often associated with hypoxemia during the period of alveolarization. In addition, exposure to excess levels of endogenous (IUGR) or exogenous (preterm birth) corticosteroids could alter lung development. Thus, the effects of IUGR and preterm birth may be additive, although this does not appear to have been studied in humans.

Early Environmental Influences on the Development of Asthma

Asthma is generally considered to be an inflammatory disorder of the conducting airways with complex underlying physiological and immunological mechanisms. The hallmark features of asthma include early and late-phase asthmatic responses (i.e. airway narrowing in response to inhaled allergens), and bronchial hyperresponsiveness (BHR). These physiological symptoms are associated with an influx into the airways of allergic-type immune cells such as eosinophils and lymphocytes. It is thought that asthma has its origins in early life and that the persistence of asthma into adulthood may be determined primarily in early childhood [45]. The best predictors of the persistence of asthma in adults are early age at asthma onset, the development of atopy to common allergens such as house dust mite (HDM), poor lung function and increased BHR in early life [46]. Other early environmental influences such as maternal smoking, mechanical ventilation of neonates and oxygen delivery to immature airways may exacerbate or even predispose to asthma, especially if the airways are exposed to these factors during late gestation and infancy when the airways are undergoing important maturation changes [47].

A systematic meta-analysis of nineteen clinical studies worldwide concluded that preterm delivery significantly increases a child's risk of developing asthma in later life [48]. However, it is not clear whether this risk is a result of the preterm birth per se or is a long-term consequence of lung injury sustained during periods of neonatal respiratory support. In very low birth weight preterm babies, mechanical ventilation is associated with BHR at 12 years of age. At this age, these children have a 2-fold increase in the risk of developing asthma compared with children born at term [49]. While BHR is a key symptom of asthma, it is interesting that very low birth weight per se was not associated with atopy [49]. This suggests that prematurity has its effects on the physiology of the disease relating to lung structure and function, rather than provoking the immune system towards an allergic bias which produces atopy and allergic inflammation. This concept is supported by a study which investigated allergen exposure in a large animal model of preterm birth [50]. In this study, adult sheep which were born 2 weeks prematurely were sensitized to HDM and then exposed to aerosols of inhaled HDM to assess allergic responses to this important allergen of humans. The study indicated that preterm birth per se, especially if followed by slow postnatal growth, was significantly associated with BHR and a prolonged late-phase bronchoconstrictor asthmatic response to allergen exposure. In contrast, atopy, as indicated by the level of IgE in serum, was not influenced by preterm birth in this model [50]. These results support the notion that preterm birth influences the physiology of the disease rather than directing the immature immune system per se towards an asthma/allergy bias.

Alterations in immune system function are believed to be closely associated with the increased prevalence of asthma observed in developed countries over the last few decades. The increased incidence of asthma and other allergic diseases in this relatively short time-frame suggests that there are strong environmental influences at play rather than any changes in the genetics of affected populations [45]. While it is not clear what the underlying causes are for this immunology-related increase in the incidence of asthma, differences in exposure to allergens and general hygiene in early life may account for some of the increases in the indices of asthma. The 'hygiene hypothesis' gives a plausible explanation for the marked increase in allergic diseases in the developed world. This hypothesis suggests that reduced exposure to microbes in early infancy deviates the immune system towards a T-helper 2 phenotype which predisposes to allergy and allergic diseases such as asthma. It is interesting that the increased predisposition for allergic asthma over the preceding decades may have coincided with an increase in the incidence of other risk factors for asthma such as preterm birth, as discussed above. The survival of preterm babies into adulthood, together with a general increase in the bias of their immune systems towards allergy would have the obvious effect of increasing the incidence of asthma. Interestingly, the prevalence of asthma in developed countries appears to have reached a plateau suggesting that the factors underlying the increased asthma rates may have peaked [51].

Conclusions

There is now a considerable body of evidence showing that exposure to a sub-optimal environment during fetal and postnatal phases of lung development can have long-term effects on lung structure, lung function and disease susceptibility. Both lung parenchyma and conducting airways are likely to be affected. Prenatal causative factors include reduced nutrient and oxygen availability, maternal tobacco smoking, exposure to high levels of corticosteroids, and postnatal factors include very preterm birth, exposure to allergens, pathogens and respiratory infections. Programming mechanisms underlying altered lung structure and function following early developmental compromises likely include permanent alterations in lung structure and epigenetic changes induced by the early developmental environment that permanently alter the expression of particular genes. Future studies should aim to identify molecular processes by which environmental factors affect lung development and whether such effects can be blocked or reversed. Ultimately however, the major goal should be to avoid a sub-optimal developmental environment through maternal education, clinical monitoring and intervention.

References

1 Minior VK, Divon MY: Fetal growth restriction at term: myth or reality? Obstet Gynecol 1998;92: 57–60.

2 Cock ML, Albuquerque CA, Joyce BJ, Hooper SB, Harding R: Effects of intrauterine growth restriction on lung liquid dynamics and lung development in fetal sheep. Am J Obstet Gynecol 2001;184:209–216.

3 Joyce BJ, Louey S, Davey MG, Cock ML, Hooper SB, Harding R: Compromised respiratory function in postnatal lambs after placental insufficiency and intrauterine growth restriction. Pediatr Res 2001;50: 641–649.

4 Dezateux C, Lum S, Hoo AF, Hawdon J, Costeloe K, Stocks J: Low birth weight for gestation and airway function in infancy: exploring the fetal origins hypothesis. Thorax 2004;59:60–66.

5 Rona RJ, Gulliford MC, Chinn S: Effects of prematurity and intrauterine growth on respiratory health and lung function in childhood. BMJ 1993;306:817–820.

6 Lawlor DA, Ebrahim S, Davey Smith G: Association of birth weight with adult lung function: findings from the British Women's Heart and Health Study and a meta-analysis. Thorax 2005;60:851–858.

7 Wignarajah D, Cock ML, Pinkerton KE, Harding R: Influence of intrauterine growth restriction on airway development in fetal and postnatal sheep. Pediatr Res 2002;51:681–688.

8 Maritz GS, Cock ML, Louey S, Suzuki K, Harding R: Fetal growth restriction has long-term effects on postnatal lung structure in sheep. Pediatr Res 2004; 55:287–295.

9 Maritz GS, Cock ML, Louey S, Joyce BJ, Albuquerque CA, Harding R: Effects of fetal growth restriction on lung development before and after birth: a morphometric analysis. Pediatr Pulmonol 2001;32:201–210.

10 Maritz G: Nicotine exposure during early development: effects on the lung; in Harding R, Pinkerton KE, Plopper CG (eds): The Lung: Development, Aging and the Environment. Amsterdam, Elsevier Academic Press, 2004, pp 301–309.

11 Harding R, Hooper SB: Regulation of lung expansion and lung growth before birth. J Appl Physiol 1996;81:209–224.

12 Barker DJ, Godfrey KM, Fall C, Osmond C, Winter PD, Shaheen SO: Relation of birth weight and childhood respiratory infection to adult lung function and death from chronic obstructive airways disease. BMJ 1991;303:671–675.

13 Bellanti JA, Zeligs BJ, Kulszycki LL: Nutrition and development of pulmonary defense mechanisms. Pediatr Pulmonol Suppl 1997;16:170–171.

14 Das RM: The effects of intermittent starvation on lung development in suckling rats. Am J Pathol 1984;117:326–332.

15 Maritz G, Probyn M, De Matteo R, Snibson K, Harding R: Lung parenchyma at maturity is influenced by postnatal growth but not by moderate preterm birth in sheep. Neonatology 2008;93:28–35.

16 Nair RH, Kesavachandran C, Shashidhar S: Spirometric impairments in undernourished children. Indian J Physiol Pharmacol 1999;43:467–473.

17 Snibson K, Harding R: Postnatal growth rate but not mild preterm birth influences airway structure in adult sheep challenged with house dust mite. Exp Lung Res 2008;93:38–35.

18 Joad JP: Tobacco smoke and lung development; in Harding R, Pinkerton KE, Plopper CG (eds): The Lung: Development, Aging and the Environment. Amsterdam, Elsevier Academic Press, 2004, pp 292–295.

19 Sekhon HS, Jia Y, Raab R, Kuryatov A, Pankow JF, Whitsett JA, et al: Prenatal nicotine increases pulmonary alpha7 nicotinic receptor expression and alters fetal lung development in monkeys. J Clin Invest 1999;103:637–647.

20 Maritz GS. Maternal nicotine exposure during gestation and lactation of rats induce microscopic emphysema in the offspring. Exp Lung Res 2002;28: 391–403.

21 Nakamura Y, Romberger DJ, Tate L, Ertl RF, Kawamoto M, Adachi Y, et al: Cigarette smoke inhibits lung fibroblast proliferation and chemotaxis. Am J Respir Crit Care Med 1995;151:1497–1503.

22 Maritz GS, Burger B: The influence of maternal nicotine exposure on neonatal lung carbohydrate metabolism. Cell Biol Int Rep 1992;16:1229–1236.

23 Paul RJ: Functional compartmentalization of oxidative and glycolytic metabolism in vascular smooth muscle. Am J Physiol 1983;244:C399–C409.

24 Contran RS, Kumar V, Robbins SL: Cellular injury and adaptation; in Pathologic Basis of Disease, ed 4. Philadelphia, WB Saunders, 1989, pp 16–38.

25 Maritz GS, Thomas RA: The influence of maternal nicotine exposure on the interalveolar septal status of neonatal rat lung. Cell Biol Int 1994;18:747–757.

26 Maritz GS: Maternal nicotine exposure and carbohydrate metabolism of fetal and neonatal lung tissue: response to nicotine withdrawal. Respiration 1987;51:232–240.

27 Windvogel S: An investigation into the effect of maternal nicotine exposure to nicotine and copper on neonatal lung development; PhD thesis, Bellville, University of Western Cape, 2007.

28 Maritz GS, Windvogel S: Is maternal copper supplementation during alveolarization protecting the developing rat lung against the adverse effects of maternal nicotine exposure? A morphometric study. Exp Lung Res 2003;29:243–260.
29 Maritz GS, Windvogel S: Does maternal nicotine exposure during different phases of lung development influence the program that regulates the maintenance of lung integrity in the offspring? Trends Comp Biochem Physiol 2005;11:63–73.
30 Tager IB, Hanrahan JP, Tosteson TD, Castile RG, Brown RW, Weiss ST, et al: Lung function, pre- and post-natal smoke exposure, and wheezing in the first year of life. Am Rev Respir Dis 1993;147:811–817.
31 Elliot J, Carroll N, Bosco M, McCrohan M, Robinson P: Increased airway responsiveness and decreased alveolar attachment points following in utero smoke exposure in the guinea pig. Am J Respir Crit Care Med 2001;163:140–144.
32 Singh SP, Barrett EG, Kalra R, Razani-Boroujerdi S, Langley RJ, Kurup V, et al: Prenatal cigarette smoke decreases lung cAMP and increases airway hyperresponsiveness. Am J Respir Crit Care Med 2003;168: 342–347.
33 Doyle LW, Faber B, Callanan C, Freezer N, Ford GW, Davis NM: Bronchopulmonary dysplasia in very low birth weight subjects and lung function in late adolescence. Pediatrics 2006;118:108–113.
34 Speer CP, Silverman M: Issues relating to children born prematurely. Eur Respir J Suppl 1998;27:13S–16S.
35 Hjalmarson O, Sandberg K: Abnormal lung function in healthy preterm infants. Am J Respir Crit Care Med 2002;165:83–87.
36 Margraf LR, Tomashefski JF Jr, Bruce MC, Dahms BB: Morphometric analysis of the lung in bronchopulmonary dysplasia. Am Rev Respir Dis 1991; 143:391–400.
37 Sweet DG, Halliday HL, Warner JA: Airway remodelling in chronic lung disease of prematurity. Paediatr Respir Rev 2002;3:140–146.
38 Albertine KH, Jones GP, Starcher BC, Bohnsack JF, Davis PL, Cho SC, et al: Chronic lung injury in preterm lambs. Disordered respiratory tract development. Am J Respir Crit Care Med 1999;159: 945–958.
39 Coalson JJ, Winter V, deLemos RA: Decreased alveolarization in baboon survivors with bronchopulmonary dysplasia. Am J Respir Crit Care Med 1995;152:640–646.
40 Bland RD, Albertine KH, Carlton DP, Kullama L, Davis P, Cho SC, et al: Chronic lung injury in preterm lambs: abnormalities of the pulmonary circulation and lung fluid balance. Pediatr Res 2000;48:64–74.
41 Sozo F, Wallace MJ, Hanna MR, Flecknoe SJ, Cock ML, Maritz GS, et al: Alveolar epithelial cell differentiation and surfactant protein expression after mild preterm birth in sheep. Pediatr Res 2006;59: 151–156.
42 Cock M, Hanna M, Sozo F, Wallace M, Yawno T, Suzuki K, et al: Pulmonary function and structure following mild preterm birth in lambs. Pediatr Pulmonol 2005;40:336–348.
43 Gilbert WM, Danielsen B: Pregnancy outcomes associated with intrauterine growth restriction. Am J Obstet Gynecol 2003;188:1596–1599.
44 Cock ML, Camm EJ, Louey S, Joyce BJ, Harding R: Postnatal outcomes in term and preterm lambs following fetal growth restriction. Clin Exp Pharmacol Physiol 2001;28:931–937.
45 Warner JO: Developmental origins of asthma and related allergic disorders; in Gluckman P, Hanson M (eds): Developmental Origins of Health and Disease. Southampton, Cambridge University Press, 2006, pp 349–369.
46 Sears MR, Greene JM, Willan AR, Wiecek EM, Taylor DR, Flannery EM, et al: A longitudinal, population-based, cohort study of childhood asthma followed to adulthood. N Engl J Med 2003;349:1414–1422.
47 Shaffer TH, Wolfson MR, Panitch HB: Airway structure, function and development in health and disease. Paediatr Anaesth 2004;14:3–14.
48 Jaakkola JJ, Ahmed P, Ieromnimon A, Goepfert P, Laiou E, Quansah R, et al: Preterm delivery and asthma: a systematic review and meta-analysis. J Allergy Clin Immunol 2006;118:823–830.
49 Mai XM, Gaddlin PO, Nilsson L, Finnstrom O, Bjorksten B, Jenmalm MC, et al: Asthma, lung function and allergy in 12-year-old children with very low birth weight: a prospective study. Pediatr Allergy Immunol 2003;14:184–192.
50 Snibson KJ, Bischof RJ, Koumoundouros E, McMurtrie LS, Cock M, Harding R: Altered airway responsiveness in adult sheep born prematurely: effects of allergen exposure. Exp Lung Res 2006;32:215–228.
51 Lugogo NL, Kraft M: Epidemiology of asthma. Clin Chest Med 2006;27:1–15.

Prof. Richard Harding
Department of Anatomy and Developmental Biology
Monash University, PO Box 13C
Clayton, Vic. 3800 (Australia)
Tel. +61 3 9905 2514, Fax +61 3 9905 2766, E-Mail richard.harding@med.monash.edu.au

Newnham JP, Ross MG (eds): Early Life Origins of Human Health and Disease.
Basel, Karger, 2009, pp 89–99

Development of the Fetal Hypothalamic-Pituitary-Adrenal-Placental Axis: Implications for Postnatal Health

John Challis[a–c] · Kristin Connor[a] · Stephen Matthews[a,c] · Stephen J. Lye[a,c] · Isabella Caniggia[c] · Felice Petraglia[e] · Alberto Imperatore[e] · Deborah M. Sloboda[g] · Shaofu Li[d] · Thorsten Braun[f] · Wei Li[a] · John P. Newnham[d]

Departments of [a]Physiology, [b]Medicine, [c]Obstetrics and Gynecology, University of Toronto, Toronto, Ont., Canada; [d]School of Women's and Infants' Health, the University of Western Australia, Perth, W.A., Australia; [e]Department of Reproductive Medicine, University of Siena, Siena, Italy; [f]Department of Obstetrics, Charité Medical University, Berlin, Germany; [g]Liggins Institute, The University of Auckland, Auckland, New Zealand

Late pregnancy is a period of stress, not only for the mother, but also for the fetus. At this time, there is an acute maturation and activation of hypothalamic-pituitary-adrenal (HPA) axis activity in the fetus with a consequent increase in the concentration of cortisol in the fetal circulation, and increased output of adrenalin from the fetal adrenal medulla. In this article, we focus on the significance of changes in fetal adrenal glucocorticoid (GC) output, in human and in sheep pregnancy, and on the relationship between endocrine activities in the fetus with the endocrine functions of the placenta.

Fetal Hypothalamic-Pituitary-Adrenal Maturation

Bassett and Thorburn [1] first reported the dramatic exponential rise in plasma cortisol that occurs in the circulation of the sheep fetus during the latter part of normal gestation. Later, we [2, 3] and others [4] showed that this was associated with a progressive increase in the concentration of adrenocorticotrophin 1–39 (ACTH) in the circulation of the fetus in a manner that was consistent with the latter driving the increase in fetal adrenal function. Fowden et al. [5] have provided a detailed review of the consistency of this change in plasma GC across animal species. In the human, as in the sheep, the major steroid secreted is cortisol, whereas in the rat and mouse the major steroid produced is corticosterone. It is evident that the rise in plasma GC provides a key component of the fetal trigger to the onset of parturition [6] and at the same time contributes to the maturation stimulus to those organ systems that the fetus requires for extrauterine survival [7, 8].

Some years ago, we described the key characteristics of the molecular changes that underlie these alterations in plasma hormone concentrations. We showed that in the fetal sheep in late pregnancy there were increases in the levels of mRNA encoding CRH and AVP in the paraventricular nucleus of the fetal hypothalamus. There was a temporal correlation with increases in levels of proopiomelanocortin (POMC) in the fetal pars distalis [9, 10]. Addition of CRH to fetal pituitary cells in culture resulted in increased accumulation of pituitary POMC mRNA and output of ACTH consistent with a causal relationship. The rise in circulating ACTH in turn upregulated expression of MC2-R (ACTH receptors) in the fetal adrenal, increased expression of key steroidogenic enzymes and enhanced coupling of the ACTH receptor through adenyl cyclase to steroidogenic enzymes. We have described these changes in detail elsewhere [10]. The acceleration in fetal HPA activity comprises one limb of the parturitional cascade that we characterized as resulting in *stimulation* of uterotonin output, with subsequent effects on a myometrium that had become *activated* through its programmed growth and proliferation cycle through pregnancy. With withdrawal of progesterone, and upregulation of the cassette of contraction proteins, including connexin-43 and stimulatory prostaglandin receptor, the myometrium is primed for the contractions that eventuate in labour.

Preterm birth occurs in 6–10% of all pregnancies [11, 12]. Its incidence has remained steady or even risen in many countries over the last 25 years. The factors contributing to preterm birth are multifactorial. In the second and third trimester of pregnancy, infection is a major cause of preterm labour. After 30 weeks' gestation, it appears that precocious activation of the fetal stress responses may play a more important role. It is clear though that pregnancies with stress as a major contributor to preterm birth may be recognized at earlier stages of gestation. Furthermore, an exaggerated myometrial inflammatory response may underlie many apparently uncomplicated deliveries both before and at term.

Thus, we have argued that activation of fetal HPA function underlies parturition at term (fig. 1). This provides an appropriate stimulus to maturation of the organ systems that the fetus requires for extrauterine survival and predicts a basis of health in later life. Conversely, some preterm labour may be associated with precocious activation of fetal HPA function, when the rise in fetal GC occurs at inappropriate times in development [13]. The fetus may develop as growth restricted, organ maturation is distorted and these small prematurely delivered babies are at great risk of predisposition to later life disease, a manifestation of the developmental origins of health and disease paradigm. We suggest that the stimuli for distorting the pattern of HPA activation may include stresses such as hypoxemia, infection and undernutrition and the mechanisms may include epigenetic processes.

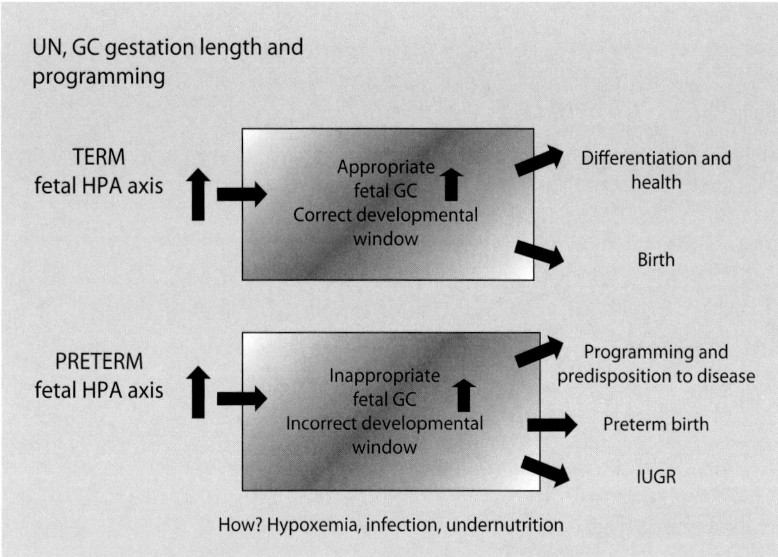

Fig. 1. Diagram depicting fetal HPA axis maturation at term, with appropriate rises in endogenous GC, and preterm, with precocious rises in GC on predisposition to later life disease and the timing of gestation length. UN = Undernutrition; IUGR = intrauterine growth restriction.

Many groups have shown that hypoxemia, generated through a variety of experimental models, results in increases in plasma ACTH and cortisol and rises in hypothalamic CRH mRNA and anterior pituitary POMC. We suggested that even small decrements of fetal arterial pO_2 were sufficient to produce transient increases in circulating ACTH that over time had the ability to increase fetal adrenal function and initiate the pathway towards birth [10].

Hypothalamic-Pituitary-Adrenal Activation, Placental CRH and Urocortins

We believe that the human fetus responds to hypoxemic stress in a similar manner to the sheep. However in the human fetus, hypoxemia increases cortisol output from the transitional zone of the fetal adrenal gland, and dehydroepiandrosterone sulphate (DHAS) from the fetal zone. DHAS can be converted to oestrogen in the placenta. Cortisol sulphate in the placenta is hydrolyzed to free cortisol which, acting through the GC receptor, upregulates expression of placental CRH. In this respect, fetal cortisol mimics the effects of stress-released maternal cortisol and counters an inhibitory effect of progesterone on placental CRH synthesis (fig. 2). We have argued that an initial effect of elevations of CRH, measurable as elevated levels in women at risk of preterm labour, may be to suppress myometrial activity through increasing cyclic

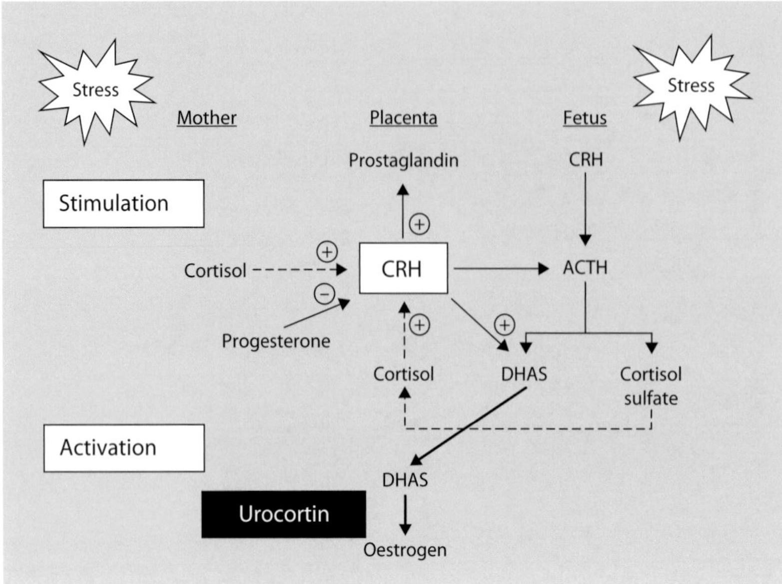

Fig. 2. Relationship between CRH, urocortin with fetal-placental oestrogen synthesis, and stimulation and activation of the uterus.

AMP. However, with sustained stimulation of this pathway, for example with prolonged hypoxemia, CRH upregulates prostaglandin-endoperoxide synthase 2 (PTGS-2) in the placenta and fetal membranes, thereby effecting the stimulus to preterm labour. Inappropriate elevations of cortisol in the fetus contribute to intrauterine growth restriction through inhibition of tissue and organ proliferation but drive further increases in placental CRH output. This might explain the association of elevated CRH levels in cord blood of women at term with a growth-restricted fetus.

Several groups of workers have shown the maternal CRH concentrations are elevated in the circulation of women at risk of later preterm delivery [6]. Furthermore, maternal CRH values are lower than normal in women subsequently delivering after term. It appears that the rises in CRH are also associated with declining levels of the circulating CRH-binding protein, thus increasing further the relative levels of free CRH in the circulation. In our own studies, we found that before 28 weeks' gestation, maternal CRH did not distinguish between women as risk of preterm birth and those not at risk. This is consistent with a different pathway, such as an augmented inflammatory response, as the primary cause of preterm birth at this time, rather than accelerated fetal HPA function.

Recently, we have explored the possibility that the CRH-related peptides, the urocortins, are also involved in these processes. Urocortin 1 has 42% amino acid homology to CRH, while urocortin 2 and 3 have 26–32% homology. Urocortin 2 and 3 are distinguished from CRH and urocortin 1 by their distinct and specific binding to the

CRH-receptor-2 (CRH-R2) species, whereas CRH and urocortin 1 bind to CRH-R1. Imperatore et al. [14] showed that urocortins 2 and 3 were expressed in gestational tissues, trophoblast, decidua and membranes throughout pregnancy. Interestingly, the levels of mRNA for urocortin 2 and 3 were higher in placenta collected at 5–8 weeks' gestation compared to tissue collected at 9–14 weeks, coinciding with the period of lower placental oxygenation. When early placental villous explants or monolayers of trophoblast cells were cultured at different oxygen tensions, the expression of urocortins 2 and 3 mRNA was lower in 8 and 3% O_2 than in cells grown at 20% oxygenation. Thus, the output of these peptides is higher at reduced oxygen tensions.

We then found that these peptides increased the conversion of C19 precursor steroids into oestrogen by placental trophoblast cells grown in vitro [15]. This activity was inhibited by an antagonist of the CRH-R2. The action on oestrogen output was accompanied by an increase in P450 aromatase mRNA in the trophoblast cells. Thus, we suggest the uncovering of a novel regulation of fetal-placental oestrogen biosynthesis in human pregnancy (fig. 2). At levels of reduced oxygen, CRH output is increased and directly stimulates the fetal zone of the fetal adrenal to produce increased amounts of oestrogen precursors. In the placenta, the related peptides, urocortin 2 and 3 are also increased by hypoxemia, and through their action in stimulating aromatase expression, they increase the conversion of C19 steroid into oestrogen. Initially, one might think of this loop as a mechanism to increase utero-placental blood flow (CRH has additional direct relaxant effects on placental blood vessels), to correct the oxygen deficiency. However, the increase in oestrogen would contribute to activation of myometrial function and CAP gene expression. Direct effects of CRH on PTGS-2 would stimulate prostaglandin biosynthesis, providing uterotonin stimulation of the activated uterus, precipitating preterm birth.

This suggestion is further complicated by the effects of oxygenation on the placental expression of the enzyme 11β hydroxysteroid dehydrogenase type 2 (11βHSD2) which interconverts cortisol to its inactive form, cortisone. 11βHSD2 is abundantly expressed in human placental syncytiotrophoblast [16] (fig. 3). It protects the fetus from exposure to inappropriate levels of maternal cortisol crossing the placenta into the fetal compartment and helps to maintain the transplacental cortisol gradient. However, placental 11βHSD2 is reduced in conditions of hypoxemia. Thus, there is less metabolism of maternal cortisol and increased cortisol availability in the placenta in addition to increased transfer across to the fetal compartment. If placental cortisol is thereby increased, one might anticipate further direct stimulation of placental CRH peptides and an exaggeration of the feed-forward loop to oestrogen biosynthesis described above. The inappropriately high levels of cortisol in the fetus contribute to programming fetal organ systems and restrained fetal growth (see below). Other conditions of lowered placental 11βHSD2 include pre-eclampsia and infection, conditions where inappropriate exposure of the fetus to elevations in maternally derived cortisol might contribute to growth restriction. Interestingly, we found that levels of mRNA encoding urocortin 2 and 3 were also elevated in the placentas of patients with pre-eclampsia [14], although at this time it is not clear whether this is a response

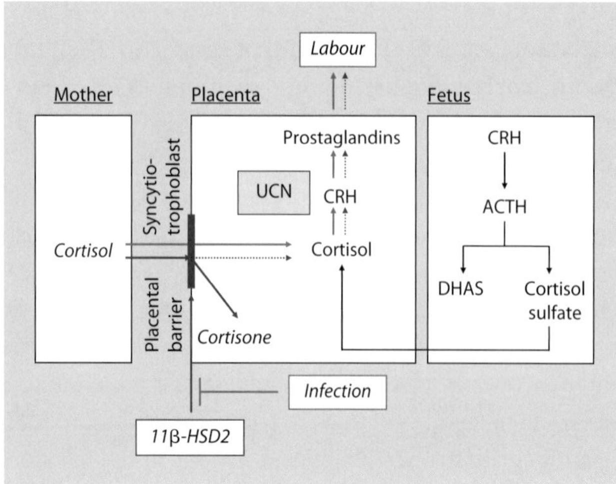

Fig. 3. Relationships between maternal and fetal HPA and placental syncytiotrophoblast 11βHSD2 in regulating the transplacental transfer and availability of cortisol during human pregnancy. UCN = Urocortin.

to reduced oxygenation or to elevations in GC or both. The effects of hypoxemia on urocortin gene expression could be reproduced at 20% oxygenation in the presence of a trap for HIF, suggesting that at least in part this is an HIF-1a-mediated pathway.

Interrelationships between Nutrition, Hypothalamic-Pituitary-Adrenal Function and Preterm Birth

Some years ago, Langley-Evans et al. [17] found that reducing the protein component of rat chow during the second half of pregnancy led to pups that were growth restricted, although average placental weights were increased. These pups developed hypertension in later life, at 16 weeks of age, consistent with an effect of altered maternal protein intake on programming later life cardiovascular function. Interestingly, the programming effect on blood pressure was abolished by maternal adrenalectomy, but was reproduced in intact control animals fed high-protein chow if placental 11βHSD2 activity was inhibited by the drug glycerinic acid, an active ingredient of liquorice. Placental 11βHSD2 was reduced in the animals fed the low protein diet, suggesting possible epigenetic regulation of this enzyme, and showing its crucial role in maintaining the appropriate balance between maternal and fetal HPA axis functions. The length of gestation was not reported in the original studies of Langley-Evans' group. However, subsequent work in Lye's laboratory has shown that a similar protocol in the mouse also reduces fetal weight, lowers placental 11βHSD2 and shortens the length of gestation [18]. These studies therefore suggested a link between maternal nutritional status and gestational length, mediated potentially through the activity of altered placental 11βHSD (fig. 3).

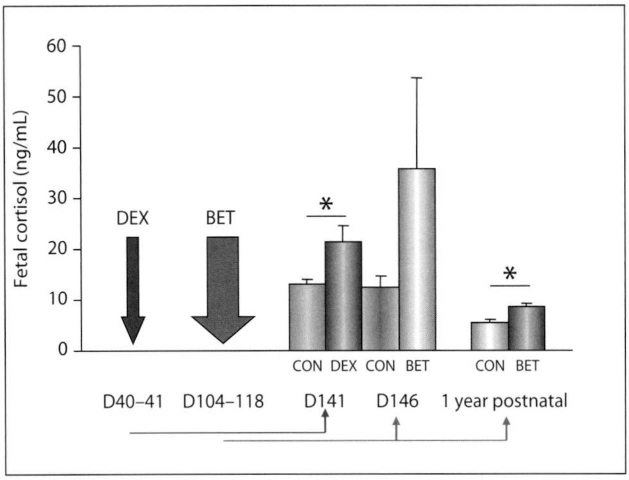

Fig. 4. Summary of basal cortisol levels in sheep after maternal treatment with exogenous synthetic GC in either early or later gestation. Basal cortisol is consistently higher in the plasma of animals where the mother had been treated with exogenous synthetic GC. For responses early in gestation, see Braun et al. [23, 24]; for responses late in gestation, see Sloboda et al. [25, 26]. * $p < 0.05$. DEX = Dexamethasone; BET = betamethasone; CON = control.

We found that sheep modestly undernourished from 60 days before the start of pregnancy to day 30 of gestation also had shortened gestation lengths [13]. In these animals, there was clear activation of fetal HPA axis activity, rises in plasma ACTH and cortisol, and increased expression of POMC in the fetal pituitary, albeit in the pars intermedia not pars distalis [19], and increased expression of P450C17 in the fetal adrenal cortex [20]. Placental 11βHSD2 activity was significantly reduced at 50 [21] and 85 days' gestational age, but not later in gestation [22]. We hypothesized that maternal peri-conceptional undernutrition had resulted in precocious activation of fetal HPA function, discernible some 100 days later in pregnancy, resulting in preterm birth in a high proportion of the animals. We speculated that this effect might be related to the reduction in placental 11βHSD2 which allowed increased transplacental transfer of maternal cortisol to the fetus. In related studies (see below), we had found that administration of a synthetic GC to maternal sheep in either early (day 39, 40) [23, 24] or later (day 104 and three times weekly thereafter) [25, 26] gestation resulted in enhanced fetal pituitary adrenal function in later gestation and in early postnatal life (fig. 4). Thus, it appeared that exposure to exogenous GC could raise the level of HPA axis function and this could contribute to the mechanisms seen after peri-conceptional undernutrition, with later increased fetal HPA activity. This thesis suggests a direct effect of maternal GC, or of increased endogenous fetal adrenal cortisol on fetal adrenal enzymatic function. In recent studies, we have followed these changes in fetal adrenal activity after administration of synthetic GC to sheep in early gestation (day 39–40) [24, 23]. Levels of ACTH in fetal cord blood were not different between control and treated animals at different stages of pregnancy, and in both groups showed the normal progressive gestation dependent rise of ACTH. At day 50, cord cortisol concentrations were lower and adrenal expression of 17α-hydroxylase (P450C17) was reduced in dexamethasone-treated animals. This indicated that maternal GC can have a direct inhibitory effect on the adrenal at the stage of pregnancy since ACTH was unchanged. There were no differences in cord cortisol or in levels of mRNA encoding ACTH receptor, steroidogenic acute regulatory protein, 3β-hydroxysteroid dehydrogenase (3βHSD) or P450C17 in the adrenals of treated and untreated animals at days 100 and 125 of pregnancy. However at term, plasma cortisol levels were significantly raised in fetuses of dexamethasone treated animals, and expression of P450C17 and 3βHSD was increased in the adrenals of female fetuses of treated mothers. Thus exposure of the fetal adrenal to excess GC in early pregnancy programs increased activity later in gestation. We found similar responses in animals whose mothers were treated with synthetic GC in later gestation (fig. 4). We suggest that the enhanced adrenal response to ACTH stimulation in human adults born with lower birth weights might reflect a relative increase in their GC exposure during intra-uterine life.

We next sought a paradigm by which we could extend these observations on effects of undernutrition to the human, to determine effects on human gestation length and enzyme changes in the placenta. The Southampton Women's Survey provided a potential source of material for study, since its subjects had been asked in a pre-pregnancy questionnaire whether they had dieted before the start of their pregnancy. However, the study had not sought information on the degree or duration of diet, or measure of change from the pre-diet BMI.

In the Southampton Women's Survey, approximately 25% of women indicated that they had dieted before or at the start of pregnancy in order to lose weight. In the first 650 patients to proceed through pregnancy, the incidence of preterm birth was higher in this group than in the group where pre- and peri-conceptional dieting had not occurred. This trend has been maintained with accumulation of information from another 1,000 women, although the relationship now just fails to achieve statistical significance. When we examined a small subset of 50 placenta from women who proceeded to term, we found that peri-conceptional dieting resulted in a significant increase in expression of PTGS-2 [27] and decrease in placental 11βHSD2 [28]. These results are consistent with our observations in the other species. Unfortunately, we do not have direct information on HPA function of the fetuses in these pregnancies. The mechanisms underlying these relationships are clearly of great importance, and are the subject of active investigation.

Effects of Exogenous Glucocorticoid on Fetal Development and Programming

The observations discussed above imply that there are adverse effects on the fetus of exposure to inappropriate levels of GC at the wrong time in gestation. Clearly, however, at term the rise in plasma cortisol is essential for the normal maturation of the lung, for glycogen deposition in the liver, for the maturation of fetal thyroid function and the switch from rT3 to T3 and for induction of enzymes of hepatic metabolism, brain and gastrointestinal function. Differentiation of these different tissues is invariably associated with a slowing in the rate of proliferation. Administration of GC at an inappropriately early time of gestation may recapitulate some of these changes, but at the expense of slowing tissue growth, leading to intrauterine growth restriction. In part, it seems likely that these effects are modulated by a reduction in the tissue-specific expression of insulin-like growth factors as stimulus for cellular proliferation and organ growth.

Ikegami et al. [29] had shown that administration to pregnant sheep of amounts of GC that mimic those used in human clinical practice in late pregnancy produced fetal growth restriction in a manner that was dependent on the amount of steroid administered. In these animals, there were reduced dimensions of femur length, abdominal circumference, and weights of the brain, liver, kidney and adrenal glands. In addition, we found that placental weights were reduced after maternal GC administration, although these had recovered by term, some 20 days after the last GC dosage. Further, we found that there was a substantial reduction in the numbers of bi-nucleate cells in the placenta [30]. These cells are the source of placental lactogen, and there was a concurrent decline in the concentration of placental lactogen in the maternal and fetal circulations. Thus, it appears that exogenous GCs, given later in gestation produce growth restriction through several different pathways. These include direct effects, presumably related to local tissue growth factor suppression, indirect effects on placental size and function, including transporter activity, and further actions through inhibiting synthesis of a metabolic hormone (oPL), that is essential for the normal provision of metabolic substrate to the placenta and transfer to the fetus.

These treatments also produced profound alteration in later life activities of different organ systems. We showed that late gestation administration of synthetic GC to the mother and/or to the fetus resulted in enhanced activity and responsiveness of the fetal pituitary-adrenal axis at term and at 1 year of age (see above) although at 3 years of age, plasma ACTH was higher and cortisol lower in the offspring of treated mothers [25, 26, 31]. Levels of mineralocorticoid receptor in the hippocampus were also raised in the offspring of treated mothers, although further studies are required to delineate the precise hippocampal regions involved in order to relate back to the changes in basal ACTH and cortisol concentrations. Several other groups have also reported sustained alterations in expression of hippocampal receptors for GC in the offspring, after maternal treatment with exogenous GC during pregnancy, implying long-term effects on the HPA axis, and on cortisol-dependent physiology, including immune responses, memory and behaviour. It is clear that these responses may involve epigenetic modification of gene activity and can be passed on to subsequent generations. It is also possible that similar mechanisms might underlie the possible changes in aggressive and attention deficit behaviours in children of mothers exposed to multiple course of corticosteroid in late pregnancy. Finally, we have reported in detail elsewhere that the offspring of maternal sheep exposed to excess GC in utero have altered insulin-glucose physiology, a dramatic increase in insulin release to a given glucose load, at least at 12 months postnatal age, and changes in expression of key enzymes of the glycogenolytic pathway [32]. Others have documented clearly that exposure of the fetus to excessive

GC in early pregnancy recapitulates many of these responses, and programs hypertension for the remainder of that individual's postnatal life.

Conclusions

In this brief chapter, we have drawn attention to the importance of pituitary-adrenal maturation in the fetus to processes of normal maturation, birth and postnatal survival. It is clear that this axis can be 'programmed' in utero. Environmental influences include the level of oxygenation, nutrition, and, not discussed here, twinning and inflammation. Likely mechanisms include changes in methylation within gene promoter regions and histone modifications. These effects may have profound influences on the outcome of gestation and on the predisposition to disease in later life.

Acknowledgements

This work was supported in large part by grants from the Canadian Institutes of Health Research, the Australian National Health and Medical Research Council and the Women and Infants Research Foundation of Western Australia. We wish particularly to thank Prof. Keith Godfrey for collaborations involving information and material collected through the Southampton Women's Survey and Prof. Peter Gluckman, Prof. Jane Harding and Dr. Frank Bloomfield for their collaboration with studies conducted in New Zealand, concerned with the effects of peri-conceptional undernutrition on HPA axis development.

References

1 Bassett JM, Thorburn GD: Foetal plasma corticosteroids and the initiation of parturition in sheep. J Endocrinol 1969;44:285–286.
2 Norman LJ, Lye SJ, Wlodek ME, Challis JRG: Changes in pituitary responses to synthetic ovine corticotrophin releasing factor in fetal sheep. Can J Physiol Pharmacol 1985;63:1398–1403.
3 Challis JR, Brooks AN: Maturation and activation of hypothalamic-pituitary-adrenal function in fetal sheep. Endocr Rev 1989;10:182–204.
4 MacIsaac RJ, Bell RJ, McDougall JG, Tregear GW, Wang X, Wintour EM: Development of the hypothalamic-pituitary-axis in the ovine fetus: ontogeny of action of ovine corticotropin-releasing factor. J Dev Physiol 1985;7:329–338.
5 Fowden, AL, Li J, Forhead AJ: Glucocorticoids and the preparation for life after birth; are there long-term consequences of the life insurance. Proc Nutr Soc 1998;57:113–122.
6 Challis JRG, Matthews SG, Gibb W, Lye SJ: Endocrine and paracrine regulation of birth at term and preterm. Endocr Rev 2000;21:514–550.
7 Liggins GC: Premature delivery of foetal lambs infused with glucocorticoids. J Endocrinol 1969;45:515–523.
8 Liggins GC: The role of cortisol in preparing the fetus for birth. Reprod Fertil Dev 1994;6:141–150.
9 Matthews SG, Han X, Lu F, Challis JRG: Developmental changes in the distribution of pro-opiomelanocortin and prolactin mRNA in the pituitary of the ovine fetus. J Mol Endocrinol 1994;13:175–185.
10 Matthews SG, Challis JRG: Regulation of the hypothalamo-pituitary-adrenocortical axis in fetal sheep. Trends Endocrinol Metab 1996;7:239–246.
11 Slattery MM, Morrison JJ: Preterm delivery. Lancet 2002;360:1489–1497.
12 Goldenberg RL, Culhane JF, Iams JD, Romero R: Epidemiology and causes of preterm birth. Lancet 2008;371:75–84.

13 Bloomfield FH, Oliver MH, Hawkins P, Campbell M, Phillips DJ, Gluckman PD, Challis JRG, Harding JE: A periconceptional nutritional origin for noninfectious preterm birth. Science 2003;300:606.

14 Imperatore A, Petraglia F, Challis JRG, Caniggia I: Urocortin 2 and urocortin 3 expression is regulated by HIF-1a in first trimester placenta explants. Reprod Sci 2007;14(suppl):432.

15 Imperatore A, Petraglia F, Challis JRG: Urocortin-2 acts through the type 2 corticotrophin releasing hormone (CRH) receptor to regulate estradiol (E2) secretion by cultured human placental trophoblasts. J Soc Gynecol Investig 2006;13(suppl):317A.

16 Alfaidy N, Gupta S, DeMarco C, Caniggia I, Challis JR: Oxygen regulation of placental 11 beta-hydroxysteroid dehydrogenase 2: physiological and pathological implications. J Clin Endocrinol Metab 2002; 87:4797–4805.

17 Langley-Evans SC, Phillips GJ, Benediktsson R, Gardner DS, Edwards CR, Jackson AA, Seckl JR: Protein intake in pregnancy, placental glucocorticoid metabolism and the programming of hypertension in the rat. Placenta 1996;17:169–172.

18 Knight BS, Pennell CE, Lye SJ: Strain differences in the impact of dietary restriction on fetal growth and pregnancy in mice. Reprod Sci 2007;14:81–90.

19 Bloomfield FH, Oliver MH, Hawkins P, Holloway AC, Campbell M, Gluckman PD, Harding JE, Challis JRG: Periconceptional undernutrition in sheep accelerates maturation of the fetal hypothalamic-pituitary-adrenal axis in late gestation. Endocrinology 2004; 145:4278–4285.

20 Connor KL, Bloomfield FH, Oliver MH, Harding JE, Challis JRG: Effect of periconceptional undernutrition on mRNA and protein expression in the fetal adrenal and placenta in late gestation. Reprod Sci 2008, submitted.

21 Jaquiery AL, Oliver MH, Bloomfield FH, Connor KL, Challis JR, Harding JE: Fetal exposure to excess glucocorticoid is unlikely to explain the effects of periconceptional undernutrition in sheep. J Physiol 2006;572:109–118.

22 Connor KL, van Zijl PL, Rumball CW, Jaquiery AL, Harding JE, Oliver MH, Bloomfield FH, Challis JRG: Do alterations in placental 11B-hydroxysteroid dehydrogenase (11BHSD) activities explain differences in fetal hypothalamic pituitary adrenal (HPA) function following periconceptional undernutrition or twinning in sheep (abstract)? Reprod Sci 2008;15: 277A.

23 Braun T, Li S, Sloboda DM, Wei L, Moss TJM, Matthews SG, Newnham JP, Challis JRG: Effects of maternal dexamethasone treatment in early pregnancy on pituitary-adrenal axis in sheep. Submitted.

24 Braun T, Li S, Matthews SG, Sloboda DM, Moss TJM, Newnham, Challis JRG: Maternal glucocorticoid injections in early pregnancy altered the pituitary-adrenal axis in sheep. Reprod Sci 2008; 15(suppl):169A.

25 Sloboda DM, Newnham JP, Challis JRG: Effects of repeated maternal betamethasone administration on growth and hypothalamic-pituitary-adrenal function of the ovine fetus at term. J Endocrinol 2000;165: 75–91.

26 Sloboda DM, Moss TJ, Gurrin LC, Newnham JP, Challis JR: The effect of prenatal betamethasone administration on postnatal ovine hypothalamic-pituitary-adrenal function. J Endocrinol 2002;172:71–81.

27 Johnstone JF, Lewis RM, Crozier S, Zelsman M, Inskip H, Hanson M, Challis JRG, Godfrey KM: The effects of parity on the duration of gestation in term pregnancies and its association with 11B HSD-1, COX-2 and PGDH protein levels in the placenta and fetal membranes (abstract). J Soc Gynecol Investig 2006;13:97A.

28 Johnstone J, Zelsman M, Crozier S, Lewis R, Inskip H, Challis JRG, Godfrey KM: The relationship between placental 11B HSD-2 and maternal body composition, age and dieting status (abstract). J Soc Gynecol Investig 2005;12:230A.

29 Ikegami M, Jobe AH, Newnham J, Polk DH, Willett KE, Sly P: Repetitive prenatal glucocorticoids improve lung function and decrease growth in preterm lambs. Am J Respir Crit Care Med 1997;156:178–184.

30 Braun T, Li S, Moss TJ, Newnham JP, Challis JRG, Gluckman PD, Sloboda D: Maternal betamethasone administration reduces binucleate cell number and placental lactogen in sheep. J Endocrinol 2007;194: 337–347.

31 Sloboda D, Moss T, Li S, Doherty D, Nitsos I, Challis JRG, Newnham J: Prenatal betamethasone exposure results in pituitary-adrenal hypo-responsiveness in adult sheep. Am J Physiol Endocrinol Metab 2006; 292:61–70.

32 Sloboda DM, Moss TJM, Li S, Doherty D, Nitsos I, Challis JRG, Newnham JP: Hepatic glucose regulation and metabolism in adult sheep: effects of prenatal betamethasone. Am J Physiol Endocrinol Metab 2005;289:E721–E728.

Dr. John R.G. Challis
Department of Physiology, University of Toronto
1 Kings College Circle, Suite 3368
Toronto, ON M5S 1A8 (Canada)
Tel. +1 416 978 4649, Fax +1 416 971 2647, E-Mail j.challis@utoronto.ca

Newnham JP, Ross MG (eds): Early Life Origins of Human Health and Disease.
Basel, Karger, 2009, pp 100–112

Developmental Origins of Musculoskeletal Disease

Cyrus Cooper[a,c] · Avan Aihie Sayer[a] · Gita Mishra[b] · Di Kuh[b]

[a]MRC Epidemiology Resource Centre, Southampton General Hospital, School of Medicine, University of Southampton, Southampton, [b]MRC Lifelong Health and Ageing Unit, University College London, London, [c]Institute of Musculoskeletal Sciences, University of Oxford, Oxford, UK

The ability to move, the protection of vital organs, and stable support for the body are the principal roles of the musculoskeletal system (muscle, bone and cartilage) [1]. This system accounts for a large proportion of the body mass; for example, the muscle mass of a healthy adult 70-kg individual is about 20 kg [2]. The musculoskeletal system develops embryonically from the mesodermal layer, differentiating into dermatomes containing skeletal and muscle cell precursors in the first trimester. At this stage, the fetus is only a few millimetres long. The growing fetus usually obtains nourishment at the expense of the mother, who tends to suffer in periods of adversity, but placental size and unrestricted blood flow through placental vessels to and from the fetus is important for optimal growth especially during the last trimester. Fetal nutrition and the uterine environment are likely to play a part in the transcription of the genomic blueprint acquired at conception into the phenotypic newborn. Some of these developmental adaptations are now known to have long-term effects on the later risk of osteoporosis and sarcopoenia. During this early phase of life, growth is rapid and there are windows of opportunity for environmental or lifestyle factors to have long-term effects, especially on the skeleton.

In this chapter, we shall review the normal patterns of bone and muscle growth, the relevance of these physiological measures to later disease, and the growing body of evidence from epidemiological and mechanistic studies that environmental influences during intrauterine and early postnatal life might lead to reduced bone mass and muscle strength in late adulthood, with a consequent increase in the risk of osteoporotic fracture.

Developmental Origins of Osteoporosis

Normal Skeletal Growth

Peak Bone Mass

At any age, the amount and quality of an individual's skeleton reflect everything that has happened from intrauterine life through the years of growth into young adulthood. The skeleton grows as the body grows, in length, breadth, mass and volumetric density. For men and women of normal body weight, total skeletal mass peaks a few years after fusion of the long bone epiphyses. The exact age at which bone mineral accumulation reaches a plateau varies with skeletal region and with how bone mass is measured. Areal density, the most commonly used measurement with dual-energy X-ray absorptiometry (DXA), peaks earliest (prior to age 20 years) at the proximal femur, while total skeletal mass peaks 6–10 years later [3].

Developmental Origins of Osteoporosis and Fracture

The importance of peak bone mass for bone strength during later life was initially suggested by cross-sectional observations that the dispersion of bone mass does not widen with age [4]. This led to the proposition that bone mass tracks throughout life and that an individual at the high end of the population distribution at age 30 years is likely to remain at that end at age 70 years. Recent longitudinal studies have confirmed this tracking, at least across the pubertal growth spurt [5].

Evidence that the risk of osteoporosis might be modified by environmental influences during early life stems from four groups of studies: (a) bone mineral measurements undertaken in cohorts of adults whose detailed birth and/or childhood records have been preserved; (b) detailed physiological studies exploring the relationship between candidate endocrine systems which might be programmed (GH/IGF-1; hypothalamic-pituitary adrenal, gonadal steroid) and age-related bone loss; (c) studies characterising the nutrition, body build and lifestyle of pregnant women and relating these to the bone mass of their newborn offspring, and (d) studies relating childhood growth rates to the later risk of hip fracture.

Epidemiological Studies

The first epidemiological evidence that osteoporosis risk might be programmed came from a study of 153 women born in Bath during 1968–1969 who were traced and studied at age 21 years [6]. Data on childhood growth were obtained from linked birth and school health records. There were statistically significant ($p < 0.05$) associations between weight at 1 year and bone mineral content (BMC), but not density, at the lumbar spine and femoral neck; these relationships were independent of adult weight and body mass index. The data suggested a discordance between the processes which govern skeletal growth, and those which influence mineralisation. They also provided direct evidence that the trajectory of bone growth might be modified in utero, an assertion previously only supported by inference from measurements of body height. The association between weight in infancy and adult bone mass was replicated in a second cohort study of 238 men and 201 women aged 60–75 years, who were born and still lived in Hertfordshire [7]. In this study, there were highly significant relationships between weight at 1 year and adult bone area at the spine and hip ($p < 0.005$); the relationships with BMC at these two sites were weaker but remained statistically significant ($p < 0.02$). They also remained after adjustment for known genetic markers of osteoporosis risk, such as polymorphisms in the gene for the vitamin D receptor [8], and after adjustment for lifestyle characteristics in adulthood which might have influenced bone mass (physical activity, dietary calcium intake, cigarette smoking, and alcohol consumption). More detailed analyses of the interactions between polymorphism in the gene for the vitamin D receptor (VDR), birthweight, and bone mineral density, have recently been published from the same cohort study [9]. In the cohort as a whole, there were no significant associations between either birthweight or VDR genotype and bone mineral density (BMD). However, the relationship between lumbar spine BMD and VDR genotype varied according to birthweight. Among individuals in the lowest third of birthweight, spine BMD was higher ($p = 0.01$) among individuals of genotype 'BB' after adjustment for age, sex and weight at baseline. In contrast, spine BMD was reduced ($p = 0.04$) in individuals of the same genotype who were in the highest third of the birthweight distribution. A statistically significant ($p = 0.02$) interaction was also found between VDR genotype and birthweight as determinants of BMD. These results suggest that genetic influences on adult bone size and mineral density may be modified by undernutrition in utero. Subsequent studies from the United States, Australia and Scandinavia have replicated these relationships between weight in infancy and adult bone mass.

Physiological Studies

To explore further the potential role of hypothalamic-pituitary function and its relevance to the pathogenesis of osteoporosis, profiles of circulating GH and cortisol were compared with bone density among groups of men and women whose birth records had been preserved. These studies revealed that birthweight and weight in infancy were predictors of basal levels of GH and cortisol during late adult life [10–12]. The levels of these two skeletally active hormones were also found to be determinants of prospectively determined bone loss rate. The data are compatible with the hypothesis that environmental stressors during intrauterine or early postnatal life alter the sensitivity of the growth plate to GH and cortisol. The consequence of such endocrine programming would be to reduce peak skeletal size, perhaps also to reduce mineralisation, and to predispose to an accelerated rate of bone loss during later life [10–12]. Recent studies suggest that interactions between the genome and early environment might establish basal levels of circulating GH, and thereby contribute to accelerated bone loss [13]. Thus, a single nucleotide polymorphism has been discovered at locus GH1-A5157G in the promoter region of the human growth hormone (GH1) gene. This is associated with significantly lower basal GH concentration, lower baseline BMD and accelerated bone loss (fig. 1). As with polymorphism in the gene for the vitamin D receptor, a significant ($p = 0.02$) interaction was observed between weight at 1 year, allelic variation at this site and bone loss rate.

Maternal Nutrition, Lifestyle and Neonatal Bone Mineral

The third piece of epidemiological evidence that osteoporosis might arise in part through developmental maladaptation stems from investigation of a series of mothers through pregnancy; anthropometric and lifestyle maternal characteristics were related to the bone mineral of their newborn offspring [14]. After adjusting for sex and gestational age, neonatal bone mass was strongly, positively associated with birthweight, birth length and placental weight. Other determinants included maternal and paternal birthweight, and maternal triceps skinfold thickness at 28 weeks. Maternal smoking and maternal energy intake at 18 weeks gestation were negatively associated with neonatal BMC at both the spine and whole body. The independent effects of maternal and paternal birthweight on fetal skeletal development support the notion that paternal influences, for example through the imprinting of growth promoting genes such as IGF-2, contribute strongly to the establishment of the early skeletal growth trajectory, while maternal nutrition and body build modify fetal nutrient supply and subsequent bone accretion, predominantly through influences on placentation.

In the most recent data from mother/offspring cohorts, body composition has been assessed by DXA in 216 children at age 9 years [15]. They and their parents had previously been included in a population-based study of maternal nutrition and fetal growth. The nutrition, body build and lifestyle of the mothers had been characterised during early and late pregnancy, and samples of umbilical venous blood had been obtained at birth. Reduced maternal height, lower pre-conceptional maternal weight, reduced maternal fat stores during late pregnancy, a history of maternal smoking and lower maternal social class were all associated with reduced whole body BMC of the child at age 9 years. Lower ionised calcium concentration in umbilical venous serum also predicted reduced childhood bone mass ($r = 0.19$, $p = 0.02$); this association appeared to mediate the effect of maternal fat stores, smoking and socio-economic status on the bone mass of the children at age 9 years. Around 25% of the mothers had sub-optimal vitamin D status as assessed by serum 25-hydroxyvitamin D concentration. The children born to these mothers had significantly ($p < 0.01$) reduced

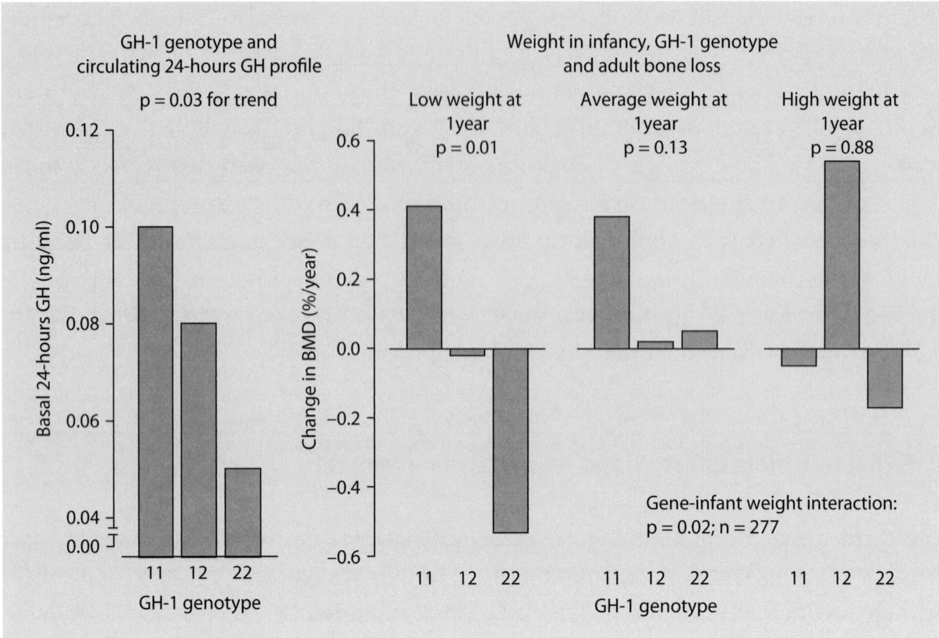

Fig. 1. GH-1 genotype, 24-hour GH concentration, weight in infancy and adult bone loss: Hertfordshire cohort study. Data derived from Dennison et al. [13].

whole body BMC at age 9 years. This deficit in skeletal growth remained significant even after adjustment for childhood weight and bone area. These data suggest that the placental capacity to maintain the materno-fetal calcium gradient is important in optimising the trajectory of postnatal skeletal growth.

Childhood Growth and Hip Fracture

Most evidence relating the intrauterine environment to later osteoporosis stems from studies utilising non-invasive assessment of bone mineral. The clinically important consequence of reduced bone mass is fracture, and data are now available which directly link growth rates in childhood with the risk of later hip fracture [16]. Studies of a unique Finnish cohort in whom birth and childhood growth data were linked to later hospital discharge records for hip fracture, have permitted follow-up of around 7,000 men and women who were born in Helsinki University Central Hospital during 1924–1933. Body size at birth was recorded and an average of 10 measurements were obtained of height and weight throughout childhood. Hip fracture incidence was assessed in this cohort using the Finnish hospital discharge registration system. After adjustment for age and sex, there were two major determinants of hip fracture risk: tall maternal height ($p < 0.001$), and low rate of childhood growth (height, $p = 0.006$; weight, $p = 0.01$; fig. 2). The effects of maternal height and childhood growth rate were statistically independent of each other, and remained after adjusting for socio-economic status. More important, hip fracture risk was also elevated ($p = 0.05$) among adults who were born short. These data are compatible with endocrine programming influencing the risk of hip fracture. In addition, the observation that fracture subjects were shorter at birth, but of average height by age 7 years, suggests that hip fracture risk might be particularly elevated among children in whom growth of the skeletal envelope is forced ahead of the capacity to mineralise, a phenomenon which is accelerated during pubertal growth.

Developmental Plasticity and Osteoporosis

Numerous animal experiments have shown that hormones, undernutrition, and other influences that affect development during sensitive periods of early life permanently programme the structure and physiology of the body's tissues and systems. A remarkable example is the effect of temperature on the sex of reptiles. If the eggs of an American alligator are incubated at 30°C, all the offspring are female. If incubated at 33°C, all the offspring are male. At temperatures between 30 and 33°C, there are varying proportions of females and males. It is believed that the fundamental sex is female, and a transcription factor is required to divert growth along a male pathway. Instead of the transcription factor being controlled genetically by a sex chromosome, it depends on the environment, specifically temperature.

Organ systems in the body are most susceptible to developmental programming during periods when they are growing rapidly. During the first 2 months of life, the embryonic period, there is extensive differentiation of progenitor cells, without rapid cell replication. Thereafter, in the fetal period, the highest growth rates are observed. Growth slows in late gestation and continues to slow in childhood. The high growth rates of the fetus compared with the child are mostly the result of cell replication; the proportion of cells which are dividing becomes progressively less as the fetus becomes older. Slowing of growth is a major adaptation to undernutrition. Experiments on rats, mice, sheep and pigs have demonstrated that protein or calorie restriction of the mother during pregnancy and lactation is associated with smaller offspring [17–20]. In general, the earlier in life that undernutrition occurs, the more likely it is to have permanent effects on body size [21]. Early in embryonic life, growth is regulated by the supply of nutrients and oxygen. At some point shortly after birth, growth begins to track. In humans, tracking is demonstrated by the way in which infants grow along centile curves. Once tracking is established, it is no longer possible to make animals grow faster by offering them unlimited food. The rate of growth has become set, homeostatically controlled by feedback systems. After a period of undernutrition, they will regain their expected size. This contrasts with the

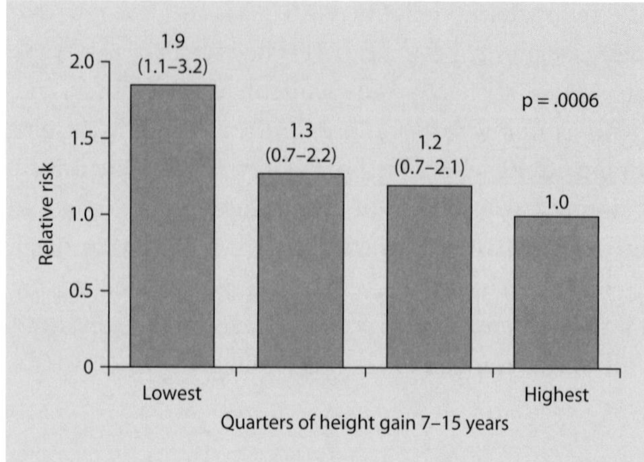

Fig. 2. Rate of childhood height gain between age 7 and 15 years, and later risk of hip fracture among 3,639 men and 3,447 women born in Helsinki University Central Hospital between 1924 and 1933.

effects of undernutrition during intrauterine life, in which skeletal development is slowed and the peak skeletal proportions attained following the completion of linear growth are reduced.

Animal models for the developmental origins of osteoporosis replicate the observations made in humans. In the first such model, the feeding of a low protein diet to pregnant rats produced offspring that exhibited a reduction in bone area and BMC, with altered growth plate morphology in adulthood [22]. Maternal protein restriction also downregulated the proliferation and differentiation of bone marrow stromal cells [23] as assessed by fibroblast colony formation at 4 and 8 weeks.

Sarcopoenia

Sarcopoenia is defined as the loss of muscle mass and strength with aging. The link between birthweight and muscle strength in older people was first observed in the Hertfordshire Ageing Study, a birth cohort study of men and women born in Hertfordshire, UK, between 1920 and 1930 and still living there 60–70 years later [24]. They had historical health visitor records of weight at birth and 1 year and were traced through the National Health Service Central Registry in Southport, UK. Following a home interview, 717 people attended a local clinic for measurement of current size and markers of aging in different body systems including grip strength. Lower birthweight and weight at 1 year were significantly associated with lower grip strength in later life, independent of adult size. This finding has now been replicated in a younger Hertfordshire cohort born 1931–1939 and in a national birth cohort of middle-aged men and women born in 1946 and participating in the National Survey of Health and Development [25]. More recent work has demonstrated a similar effect size of birthweight on adult muscle strength in young women aged 20–34 years taking part in the Southampton Women's Survey, suggesting an association between early size and peak muscle strength rather than decline [26].

It has been possible to add to these findings with a more detailed life course approach using longitudinal data collected in the National Survey of Health and Development [27]. Grip strength and body size were measured in a representative British sample of 1,406 men and 1,444 women who were 53 years old and had prospective childhood data on weight, height, motor milestones, cognitive ability and information on lifetime social class, current physical activity and health status. Birthweight and pre-pubertal height gain were associated with midlife grip strength, independently of later weight and height gain (fig. 3). Pubertal growth was also independently associated with midlife grip strength; for men, weight gain during puberty was beneficial, whereas for women it was height gain. Those participants with earlier infant motor development had better midlife grip strength, which was partly confounded by the growth trajectory. This suggests that components of prenatal, prepubertal, and pubertal growth have long-term effects on midlife grip strength [28].

Studies investigating the relationship between growth in early life and muscle mass have demonstrated consistent findings linking low birthweight with reduced muscle mass. A study of older men participating in the Hertfordshire Cohort Study showed that birthweight was significantly positively associated with fat-free mass but not with measures of adult fat mass. In contrast, weight at 1 year was associated with fat-free mass and adult fat mass estimated using anthropometry [29]. Similar findings were observed in studies of men and women using urinary creatinine excretion [30] and DXA [31] to estimate muscle mass. More recently, a study on the Helsinki birth cohort in Finland has replicated the relationships between small size at birth, lower muscle mass and reduced grip strength in older people [32]. Studies of birthweight and muscle mass in earlier stages of life demonstrate similar findings for children [33], teenagers [34] and young adults [35].

Nutrition

There is considerable interest in which aspects of the early environment underlie these associations. Few retrospective cohort studies have sufficiently detailed data on prenatal and postnatal environmental influences to assess specific effects on long-term muscle mass and strength, but these questions are being addressed in the prospective Southampton Women's Survey [36]. This study has recruited over 12,500 women living in the city of Southampton and interviewed them to assess health, body composition, lifestyle and diet. They have been followed in subsequent pregnancies and their offspring followed through childhood, with the aim of identifying prospectively the influence of the pre-conceptional and antenatal environment on the growth and development of the fetus, infant and child.

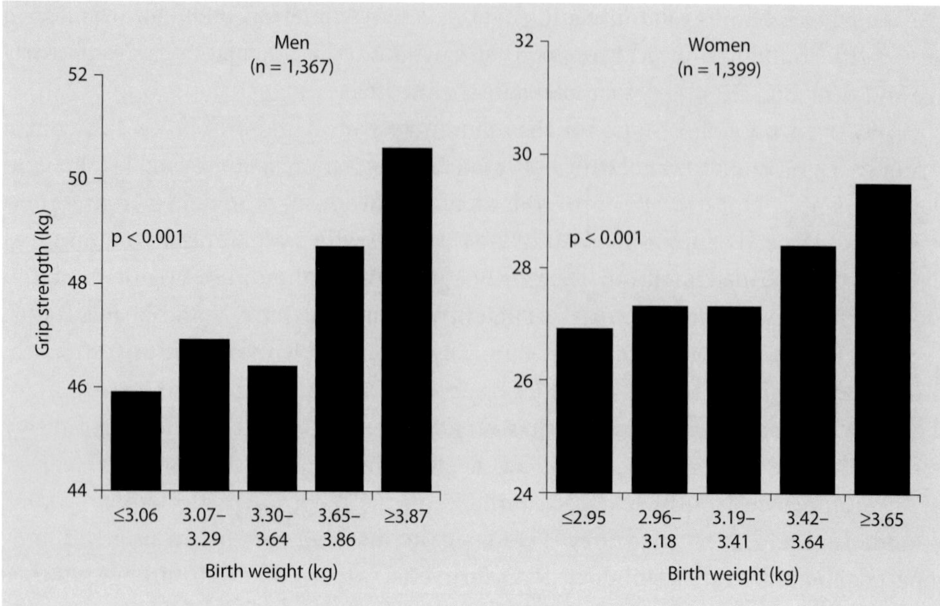

Fig. 3. Grip strength at 53 years according to birthweight in the National Survey of Health and Development 1946 birth cohort [28].

We have used data from 448 mother-offspring pairs in this cohort to examine parental influences on neonatal body composition, ascertained by DXA [37]. Taller women and those with higher parity had offspring with increased birthweight, fat and lean mass, whereas women who smoked during pregnancy had smaller babies, with reduced fat and lean mass. Maternal walking speed was negatively associated with birthweight, and fat mass positively predicted neonatal total and proportionate fat but was negatively correlated with proportionate lean mass. Future analyses will focus on the influence of maternal diet on neonatal body composition and childhood grip strength.

There is evidence from animal models that prenatal undernutrition is associated with reduced neonatal muscle weight in sheep [38], and a reduction in postnatal muscle fibre number in the pig [39], guinea pig [40], and rat [41]. There is evidence that these effects persist. This contrasts with the beneficial effects on aging of dietary restriction instituted later in life [42]. Muscle fibre number is a critical determinant of muscle mass and strength, and a number of studies have described the regulators of myofibre number, type and size [43]. Genetic factors appear to be the major influence on primary fibre number, whereas environmental factors such as maternal undernutrition have a predominant effect on the growth and development of secondary fibres [44]. This evidence suggests that early influences on muscle fibres may underlie the epidemiological associations between size at birth and adult muscle mass and strength.

However, the concept that there is a fixed number of muscle fibres determined by birth, with subsequent growth only achieved by increase in fibre size, is probably now outdated. Recent evidence suggests that post-mitotic myonuclei lying within mature myofibres might be able to reform myoblasts or stem cells, and there is increasing recognition of the role that satellite cells play in postnatal muscle growth and regeneration [45, 46]. One small clinical study has looked at the relationship between size at birth and adult human skeletal muscle morphology, but did not find significant associations. Current work is focused on further elucidating the molecular and cellular mechanisms of developmental influences on sarcopoenia.

The evidence linking nutrition and muscle strength in healthy individuals across the life course is surprisingly limited. Dietary protein may be important and protein intake may be insufficient with the reduction in food intake seen in later life. There is concern that the dietary reference intake for protein may be set too low to ensure optimal intakes in healthy elderly adults; however, protein supplementation studies are not always successful in influencing muscle function [47]. More consistent links have been demonstrated between vitamin D status and intake, and markers of muscle function. Beneficial effects of improved status have been described in both observational and supplementation studies [48, 49]. There is also some interest in the role of antioxidant status in the development of sarcopoenia because of its possible link with oxidative stress [50], although it is not known how variation in intakes of antioxidant nutrients impact on muscle function or rate of muscle loss.

Physical Activity

There is consistent evidence linking level of physical activity with muscle mass and strength in later life [51, 52], and in particular resistance exercise training is the most effective intervention to ameliorate the loss of muscle mass and strength with age [53, 54]. However, it is recognized that sarcopoenia is not solely due to reduced levels of physical activity in older people, as some loss of muscle mass and strength is experienced even by elite athletes maintaining very high levels of exercise into later life [55]. The relative contribution of levels of physical activity across different stages of the life course to long-term muscle mass and strength is far less clear, and the potential for exercise interventions earlier in life to modify sarcopoenia needs to be explored.

Conclusion

UK birth cohorts such as the Hertfordshire Cohort Study and the National Survey of Health and Development with longitudinal measurements of growth, diet, physical activity and lifestyle offer the opportunity to investigate both the individual contributions of different environmental exposures over time and the interrelationships between them. This approach has the potential to allow investigation of the complex influences on the aging muscle phenotype and inform the development of effective interventions in aging and sarcopoenia across the life course.

Acknowledgements

We are grateful to the Medical Research Council, the Wellcome Trust, the Arthritis Research Campaign, the National Osteoporosis Society and the British Heart Foundation for support of our research programme into the developmental origins of osteoporotic fracture. The manuscript was prepared by Sue Curtis.

References

1 Simkin PA: The musculoskeletal system; in Klippel JH, Dieppe PA (eds): Rheumatology. London, Mosby, 1994, pp 1.2.1–1.2.10.

2 Kreisberg RA, Bowdoin B, Meador CK: Measurement of muscle mass in humans by isotopic dilution of creatine 14C. J Appl Physiol 1970;28:264–267.

3 Matkovic V, Jelic T, Wardlaw GM, Ilich JZ, Goel PK, Wright JK, et al: Timing of peak bone mass in Caucasian females and its implication for the prevention of osteoporosis. J Clin Invest 1994;93:799–808.

4 Newton JHF, Morgan BD: The loss of bone with age: osteoporosis and fractures. Clin Orthop 1970;71: 229–232.

5 Ferrari S, Rizzoli R, Slosman D, Bonjour JP: Familial resemblance for bone mineral mass is expressed before puberty. J Clin Endocrinol Metab 1998;83: 358–361.

6 Cooper C, Cawley MID, Bhalla A, Egger P, Ring F, Morton L, Barker D: Childhood growth, physical activity and peak bone mass in women. J Bone Miner Res 1995;10:940–947.

7 Cooper C, Fall C, Egger P, Hobbs R, Eastell R, Barker D: Growth in infancy and bone mass in later life. Ann Rheum Dis 1997;56:17–21.

8 Keen R, Egger P, Fall C, Major P, Lanchbury J, Spector TD, Cooper C: Polymorphisms of the vitamin D receptor, infant growth and adult bone mass. Calcif Tiss Int 1997;60:233–235.

9 Dennison EM, Arden NK, Keen RW, Syddall H, Day INM, Spector TD, Cooper C: Birthweight, vitamin D receptor genotype and the programming of osteoporosis. Paediatr Perinat Epidemiol 2001;15:211–219.

10 Fall C, Hindmarsh P, Dennison E, Kellingray S, Barker D, Cooper C: Programming of growth hormone secretion and bone mineral density in elderly men; an hypothesis. J Clin Endocrinol Metab 1998; 83:135–139.

11 Dennison E, Hindmarsh P, Fall C, Kellingray S, Barker D, Phillips D, Cooper C: Profiles of endogenous circulating cortisol and bone mineral density in healthy elderly men. J Clin Endocrinol Metab 1999;84:3058–3063.

12 Phillips DIW, Barker DJP, Fall CHD, et al: Elevated plasma cortisol concentrations: a link between low birthweight and the insulin resistance syndrome? J Clin Endocrinol Metab 1998;83:757–760.

13 Dennison EM, Syddall HE, Rodriguez S, Voropanov A, Day INM, Cooper C, the Southampton Genetic Epidemiology Research Group: Polymorphism in the growth hormone gene, weight in infancy, and adult bone mass. J Clin Endocrinol Metab 2004;89: 4898–4903.

14 Godfrey K, Walker-Bone K, Robinson S, Taylor P, Shore S, Wheeler T, Cooper C: Neonatal bone mass: influence of parental birthweight, maternal smoking, body composition, and activity during pregnancy. J Bone Miner Res 2001;16:1694–1703.

15 Javaid MK, Crozier SR, Harvey NC, Gale CR, Dennison EM, Boucher BJ, Arden NK, Godfrey KM, Cooper C, Princess Anne Hospital Study Group: Maternal vitamin D status during pregnancy and childhood bone mass at age 9 years: a longitudinal study. Lancet 2006;367:36–43.

16 Cooper C, Eriksson JG, Forsén T, Osmond C, Tuomilehto J, Barker DJP: Maternal height, childhood growth and risk of hip fracture in later life: a longitudinal study. Osteoporosis Int 2001;12:623–629.

17 Barker DJP: Programming the baby; in Barker DJP (ed): Mothers, Babies and Health in Later Life. London, Churchill Livingstone, 1998, pp 13–41.

18 McCance RA, Widdowson EM: The determinants of growth and form. Proc R Soc Lond B 1974;185:1–17.

19 Smart JL, Massey RF, Nash SC, Tonkiss J: Effects of early life undernutrition in artificially reared rats: subsequent body and organ growth. Br J Nutr 1987; 58:245–255.
20 Chow BF, Lee CJ: Effect of dietary restriction of pregnant rats on body weight gain of the offspring. J Nutr 1964;82:10–18.
21 McCance RA, Widdowson EM: Nutrition and growth. Proc R Soc Lond B 1962;156: 326–337.
22 Mehta G, Roach HI, Langley-Evans S, Taylor P, Reading I, Oreffo ROC, Aihie Sayer A, Clarke NMP, Cooper C: Intrauterine exposure to a maternal low protein diet reduces adult bone mass and alters growth plate morphology in rats. Calcif Tiss Int 2002;71:493–498.
23 Oreffo ROC, Lashbrooke B, Roach HI, Clarke NMP, Cooper C: Maternal protein deficiency affects mesenchymal stem cell activity in the developing offspring. Bone 2003;33:100–107.
24 Aihie Sayer A, Cooper C, Evans JR, et al: Are rates of aging determined in utero? Age Ageing 1998;27: 579–583.
25 Aihie-Sayer A, Syddall HE, Gilbody HJ, Dennison EM, Cooper C: Does sarcopenia originate in early life? Findings from the Hertfordshire cohort study. J Gerontol A Biol Sci Med Sci 2004;59:M930–M934.
26 Inskip HM, Godfrey KM, Martin HJ, Simmonds SJ, Cooper C, Sayer AA, Southampton Women's Survey Study Group: Size at birth and its relation to muscle strength in young adult women. J Intern Med 2007;262:368–374.
27 Wadsworth M, Kuh D, Richards M, Hardy R: Cohort Profile: The 1946 National Birth Cohort (MRC National Survey of Health and Development). Int J Epidemiol 2006;35:49–54.
28 Kuh D, Hardy R, Butterworth S, Okell L, Richards M, Wadsworth M, Cooper C, Sayer AA: Developmental origins of midlife physical performance: evidence from a British birth cohort. Am J Epidemiol 2006;164: 110–121.
29 Sayer AA, Syddall HE, Dennison EM, Gilbody HJ, Duggleby SL, Cooper C, Barker DJ, Phillips DI: Birth weight, weight at 1 year of age, and body composition in older men: findings from the Hertfordshire Cohort Study. Am J Clin Nutr 2004;80:199–203.
30 Phillips DIW: Relation of fetal growth to adult muscle mass and glucose tolerance Diabet Med 1995;12:686–690.
31 Gale CR, Martyn CN, Kellingray S, Eastell R, Cooper C: Intrauterine programming of adult body composition. J Clin Endocrinol Metab 2001;86:267–272.
32 Ylihärsilä H, Kajantie E, Osmond C, Forsén T, Barker DJ, Eriksson JG: Birth size, adult body composition and muscle strength in later life. Int J Obes 2007;31:1392–1399.
33 Hediger ML, Overpeck MD, Kuczmarski RJ, McGlynn A, Maurer KR, Davis WW: Muscularity and fatness of infants and young children born small or large for gestational age. Pediatrics 1998;102:E60.
34 Singhal A, Wells J, Cole TJ, Fewtrell M, Lucas A: Programming of lean body mass: a link between birth weight, obesity, and cardiovascular disease? Am J Clin Nutr 2003;77:726–730.
35 Khan HS, Narayan KM, Williamson DF, Valdez R: Relation of birth weight to lean and fat thigh tissue in young men. Int J Obes Relat Metab Disord 2000; 24:667–672.
36 Inskip HM, Godfrey KM, Robinson SM, Law CM, Barker DJ, Cooper C, SWS Study Group: Cohort profile: The Southampton Women's Survey. Int J Epidemiol 2006;35:42–48.
37 Harvey NC, Poole JR, Javaid MK, Dennison EM, Robinson S, Inskip HM, Godfrey KM, Cooper C, Sayer AA, SWS Study Group: Parental determinants of neonatal body composition. J Clin Endocrinol Metab 2007;92:523–526.
38 Greenwood PL, Slepetis RM, Bell AW: Influences on fetal and placental weights during mid to late gestation in prolific ewes well nourished throughout pregnancy. Reprod Fertil Dev 2000;12:149–156.
39 Dwyer CM, Stickland NC, Fletcher JM: The influence of maternal nutrition on muscle fiber number development in the porcine fetus and on subsequent postnatal growth. J Animal Sci 1994;72:911–917.
40 Dwyer CM, Madgwick AJ, Ward SS, Stickland NC: Effect of maternal undernutrition in early gestation on the development of fetal myofibres in the guinea-pig. Reprod Fertil Dev 1995;7:1285–1292.
41 Wilson SJ, Ross JJ, Harris AJ: A critical period for formation of secondary myotubes defined by prenatal undernourishment in rats. Development 1988; 102:815–821.
42 Prakash YS, Fournier M, Sieck GC: Effects of prenatal undernutrition on developing rat diaphragm. J Appl Physiol 1993;75:1044–1052.
43 Dauncey MJ, Gilmour RS: Regulatory factors in the control of muscle development. Proc Nutr Soc 1996; 55:543–559.
44 Maltin CA, Delday MI, Sinclair KD, Steven J, Sneddon AA: Impact of manipulations of myogenesis in utero on the performance of adult skeletal muscle. Reproduction 2001;122:359–374.
45 Bailey P, Holowacz T, Lassar AB: The origin of skeletal muscle stem cells in the embryo and the adult. Curr Opin Cell Biol 2001;13:679–689.

46 Stewart CE: The physiology of stem cells: potential for the elderly patient. J Musculoskelet Neuronal Interact 2004;4:179–183.

47 Thompson CH, Sanderson AL, Sandeman D, Stein C, Borthwick A, Radda GK, Phillips DI: Fetal growth and insulin resistance in adult life: role of skeletal muscle morphology. Clin Sci 1997;92:291–296.

48 Janssen HC, Samson MM, Verhaar HJ: Vitamin D deficiency, muscle function, and falls in elderly people. Am J Clin Nutr 2002;75:611–615.

49 Boland R: Role of vitamin D in skeletal muscle function. Endocr Rev 1986;7:434–448.

50 Weindruch R: Interventions based on the possibility that oxidative stress contributes to sarcopenia. J Gerontol A Bio Sci Med Sci 1995;50:157–161.

51 Brach JS, Simonsick EM, Kritchevsky S, Yaffe K, Newman AB, Health, Aging and Body Composition Study Research Group: The association between physical function and lifestyle activity and exercise in the health, aging and body composition study. J Am Geriatr Soc 2004;52:502–509.

52 Bassey EJ: Measurement of muscle strength and power. Muscle Nerve Suppl 1997;5:S44–S46.

53 Baker MK, Atlantis E, Fiatarone Singh MA: Multimodal exercise programs for older adults. Age Ageing 2007;36:375–381.

54 Borst SE: Interventions for sarcopenia and muscle weakness in older people. Age Ageing 2004;33:548–555.

55 Hawkins SA, Wiswell RA, Marcell TJ: Exercise and the master athlete – a model of successful aging? J Gerontol A Biol Sci Med Sci 2003;58:1009–1011.

Prof. Cyrus Cooper
MRC Epidemiology Resource Centre
Southampton General Hospital
Southampton SO16 6YD (UK)Tel. +44 23 8077 7624, Fax +44 23 8070 4021, E-Mail cc@mrc.soton.ac.uk

Newnham JP, Ross MG (eds): Early Life Origins of Human Health and Disease.
Basel, Karger, 2009, pp 113–122

Developmental Origins of Reproductive Health

D.M. Sloboda[a] · R. Hart[b] · M. Hickey[b,c]

[a]Liggins Institute, The University of Auckland, Auckland, New Zealand; [b]School of Women's and Infants Health, The University of Western Australia, [c]Women and Infants Research Foundation, King Edward Memorial Hospital, Perth, W.A., Australia

There is now compelling evidence that adult disease risk is modified by events that occur early in life, involving interactions between the genome and the developmental environment. Data from extensive epidemiological, prospective clinical and experimental studies have given rise to the concept of developmental programming, whereby an unfavourable early environment is believed to trigger adaptations that improve fetal and/or infant survival and prepare the individual for a particular range of environments later in life [1]. These processes, while adaptive in their prior evolution, may subsequently prove to be maladaptative. This may be particularly true when the pre- and postnatal environments are widely discrepant. Evidence that reproductive function may be influenced by intrauterine events is emerging from animal studies and selected human populations [2, 3], joining a growing list of 'adult-onset' diseases including obesity and type 2 diabetes that may have their origins in utero. This chapter will overview the evidence that early life events affect reproductive health in adolescence and adulthood.

Developmental Origins of Female Reproductive Health

Developmental Factors Regulating Pubertal Development

The onset of puberty is the first indicator of emerging reproductive function, and many studies support associations between early life growth and pubertal development. Growth during childhood is associated with age at puberty in both boys and girls [4, 5]. Fetal growth has been correlated with the onset of adrenarche (maturation of the adrenal cortex related to puberty but distinct from hypothalamic-pituitary-gonadal maturation and function) [6]. Longitudinal cohort studies of growth-restricted infants suggest that fetal growth restriction is associated with early-onset adrenarche, pubarche, ovarian hyperandrogenism and hyperinsulinemia during adolescence [7–10]. Opposing influences between prenatal growth restriction and accelerated childhood weight gain have been associated with premature adrenarche, menarche, hyperinsulinemia, dyslipidemia, and obesity during childhood as well as polycystic ovary syndrome (PCOS) in adolescence [11–14].

Developmental Factors Regulating Onset of Menarche

Menarche is the first indicator of female reproductive capacity and the timing of menarche is influenced by a combination of prenatal factors and postnatal growth. The past century has seen a dramatic decrease in the age at menarche; in European girls, the age of menarche has fallen from 17 to ~12.5 years of age in the last 100 years [2, 15]. This trend is likely due to the dramatic improvement in child health since the early 19th century when the age of menarche was highest. Earlier age at menarche may have adverse consequences in adolescence and adult life; it is a risk factor for teenage depression [16], insulin resistance [17, 18] and breast cancer in adulthood [19] and is linked with subsequent obesity [20]. The recent increase in childhood obesity may be a contributing factor in the age of onset of puberty and menarche, although mechanisms are poorly understood. In early studies, early life weight gain was associated with age at menarche [21], and since then reports across a number of different populations have confirmed that prenatal and postnatal growth can independently predict age at menarche. We have shown that in girls, low birthweight followed by accelerated weight gain in childhood is associated with earlier menarche [22] (table 1). In this cohort of otherwise unremarkable adolescents, girls that were below the median estimated birthweight and above the median in body mass index at childhood (age 8 years) entered puberty significantly earlier than girls that were relatively heavy at birth and leaner at age 8 [22]. Low birthweight girls with catch-up growth have earlier menarche [8, 23], reduced ovarian size [10], reduced ovulation rate [24] and ovarian hyperandrogenism after precocious pubarche [7]. Prenatal growth restriction is also associated with alterations in gonadotropin levels (FSH, inhibin) in infancy [25] and in adolescents [10]. Finally, poor growth and development in late gestation may be associated with earlier cessation of menstruation and the onset of menopause [26], highlighting the potential for long-lasting consequences.

The mechanisms regulating these associations in human studies are unclear, but may relate to insulin sensitivity. In low birthweight girls, insulin sensitization with metformin decreased circulating leptin and insulin-like growth factor 1 levels and resulted in a significant delay of menarcheal onset [27], where improvement in their metabolic profile still exists 4 years after treatment [28]. It is therefore likely that a disruption of the adipoinsular axis may contribute significantly to the association between early growth restriction, postnatal adiposity and pubertal development, but this has yet to be confirmed empirically.

Table 1. Age at menarche stratified by EBW and BMI at 8 years of age

EBW and BMI subgroup	Cohort size	Girls who reached menarche		Median age	IQ range	Range
		n	%			
EBW <1 and BMI ≥16.3	231	120	52	12.5	12.1–13.2	9.4–14.4
EBW ≥1 and BMI ≥16.3	306	142	46	12.8	12.2–13.6	9.8–14.6
EBW <1 and BMI <16.3	127	54	43	13.0	12.6–14.2	10.6–14.6
EBW ≥1 and BMI <16.3	112	33	29	13.2	12.8–14.4	11.0–14.2
Total	776	349				

Adapted from Sloboda et al. [22]. EBW = Expected birthweight ratio; BMI = body mass index. EBW and BMI are illustrated as less than or greater than/equal to the median value for the whole cohort.

Early age at menarche (or puberty) may be the interaction of early life events to advance menarche under poor conditions and the childhood influence to advance menarche under enriched conditions where both events are designed to advance menarche/puberty. This phenomenon has been demonstrated in populations migrating from a poor developing country to a developed country, primarily through international adoption [for review, see 29]. Recently, a large registry-based study of adopted children in Denmark has demonstrated that adopted girls from India and South America were 10–20 times more likely to develop precocious puberty than the Danish reference group [30] confirming findings found in smaller cohort studies [31]. Young adopted girls (between 5–8 years of age) in Denmark also showed signs of increased pituitary and gonadal activity despite their prepubertal phenotype, suggesting that early onset of puberty in these adopted girls may be centrally driven [32].

Developmental Origins of the Polycystic Ovarian Syndrome

The PCOS is the most common reproductive disorder in women, affecting around 6% of the adult population [33]. PCOS is associated with menstrual irregularity and ovulatory disorder in conjunction with hyperandrogenism, and a classical appearance of cystic ovaries in PCOS is often associated with insulin resistance, obesity and dislipidemia [34].

Prenatal growth patterns have been associated with a prevalence of PCOS in girls, and girls of low birthweight are at high risk of PCOS, particularly those with a history of precocious pubarche [35]. Animal models of PCOS have focussed on the role of early life androgen exposure. PCOS models have been successfully developed in sheep [36] and non-human primates [37] through intrauterine exposure to androgen. These studies have shown that prenatal androgen exposure results in offspring with enlarged cystic ovaries with no corporea lutea, anovulation, raised gonadotropin (luteinising hormone, LH) levels, progressive loss of ability to develop an LH surge in response to increasing estradiol levels, ovarian hyperandrogenism and central obesity and insulin resistance. It is these features that are commonly found in the adult phenotype of PCOS in humans [38]. In women born with congenital adrenal hyperplasia or early virilising tumours, some features of PCOS are manifest such as a raised basal LH level, an increased responsiveness of LH and 17-hydroxyprogesterone to gonadotropin-releasing hormone infusion, confirming the hypothesis that exposure to elevated androgens is an important contributing factor in PCOS. In pregnant women with PCOS, androgen levels peak at 22–28 weeks [39], highlighting the potential for PCOS mothers to contribute to the next generation's PCOS phenotype downstream of increased exposure of the female fetus to maternally derived androgens. Under normal circumstances, placental aromatase enzyme provides a protective barrier against fetal exposure to maternal androgens; however, it is possible that during times of placental insufficiency or aromatase deficiency the fetus may be exposed to higher levels of androgens, although it has yet to be proven.

Developmental Origins of Male Reproductive Health

In boys, the onset of reproductive capability is a dynamic ongoing process, without the clear watershed marker of menarche, and as a result data describing pubertal onset in boys are less well established than those presented for girls. Nevertheless, emerging data suggest that early life events also influence male reproductive health [40, 41]. Evidence exists to suggest that early life events may contribute to male infertility. Approximately 1 in 6 couples suffers from involuntary subfertility; generally defined as being unable to conceive after 1 year of trying [42] and male factors contribute to approximately 33–50% of these cases [43]. Men whose mothers smoked during pregnancy demonstrated decreased adult sperm cell counts compared with those whose mothers did not smoke [41], suggestive of testicular spermatogenic vulnerability to early life events. Low birthweight has been associated with male subfertility [44]. It has been demonstrated that adult men born short for gestational age also have higher gonadotropin drive coincident with lower testosterone and inhibin B levels, suggesting poor testicular responsiveness potentially influencing fertility [45].

Mechanistic Factors Underlying the Programming of Reproductive Health

Early Life Nutrition

There is no doubt that nutrition during childhood can influence the timing of puberty [15], where poor childhood nutrition delays puberty and good childhood nutrition accelerates pubertal maturation [46]. Intrauterine undernutrition and low birth-weight, conversely, accelerate menarche [21, 47]. The best datasets suggestive of an effect of interaction between prenatal and postnatal environments on reproductive maturation are those of international adoption studies. As in the aforementioned section, girls from impoverished countries such as India and Bangladesh adopted into relatively wealthy countries such as Denmark, demonstrate early-onset menarche [30, 48]. If girls migrated with their families, this effect was not seen; firm suggestion of an environmental (or nutritional) mismatch in those girls that migrated alone and were adopted into Danish families [30].

Much of what we know regarding nutrition and postnatal reproductive outcomes comes from experimental studies. In the rat, it is well established that postnatal fat (or weight gain) is a primer for pubertal onset; classic studies have shown that female rats fed a high-fat diet after weaning demonstrate early-onset puberty [49, 50]. Over the last decade, however, studies have shifted focus to the intrauterine environment. Maternal protein restriction throughout pregnancy in the rat had no significant effect on pubertal age in male and female offspring but significantly reduced long-term fertility rate [51] as well as longevity in adult females [52]. Conversely, offspring of pregnant rats fed a high-fat diet enter puberty earlier and have their first oestrus earlier compared with control offspring [53].

Endocrine Disruptors

Environmental toxins have the potential to influence reproductive function [54], and evidence from both experimental models and human population studies demonstrates that exposure to hormonally active environmental toxicants, also called endocrine disruptors, alter reproductive development and increase the risk of reproductive disorders [55–57].

There are, however, limitations with these studies; a wide range of potential endocrine disruptors exist in the environment, difficulty in measuring toxin exposure, potential cumulative effects of endocrine disruptors and the extent to which the fetus has been affected by maternal exposure. The effect of endocrine disruptors on subsequent reproductive health clearly depends on critical periods or windows of vulnerability.

Comprehensive reviews of studies investigating the effects of endocrine disruptors and pubertal onset have highlighted the fact that environmental toxins are widely variant and may have either oestrogenic, anti-oestrogenic or anti-androgenic properties, confusing reported effects [57]. In general, those that have oestrogenic properties have been shown to advance puberty maturation, although there are several studies that report no association with exposure to endocrine disruptors [58]. Many endocrine disruptors have lipophilic properties that enable them to sequester in lipid-rich tissue, making them potentially active for many years even after initial exposure. Dose-response relationships are at best difficult to measure and one should never assume that all endocrine disruptors have a similar dose-response relationship, as these compounds show different mechanisms of action [57].

Endocrine disruptors that have oestrogenic qualities are of particular interest in reproductive health since they can modulate oestrogenic signalling [59]. A positive association between precocious puberty and in utero exposure to environmental chemicals has been demonstrated [60, 61], although there exist studies that have found no effect [62, 63] and even a delay in puberty [64]. Other reproductive dysfunctions have also been attributed to prenatal exposure to endocrine disruptors; adolescent girls born to mothers exposed to polychlorinated biphenyls/polychlorinated dibenzofurans during pregnancy showed menstrual irregularity and altered gonadotropin function [65], and polybrominated diphenyl ethers commonly used as flame retardants, have been found to exist in the placenta and breast milk of mothers whose newborn boys demonstrated cryptorchidism and testicular dysfunction [66].

Epigenetic Regulation of Early Life Programming

At a mechanistic level, much attention is now focused on the epigenetic regulation of gene expression, and it has been proposed that DNA and chromatin modifications during vulnerable windows of development might underpin much of the disease risk associated with early life influences. Epigenetics is the study of mitotically and/or meiotically heritable changes in gene function that cannot be explained by changes in DNA sequence [67]. Whereas genetic inheritance involves sequence changes (e.g. mutation, random assortment), epigenetic processes lead to heritable changes in gene function by altering DNA chemistry independent of sequence. This has the effect of causing cells with identical genetic information to display different phenotypes. There is increasing interest in the role that internal and external environmental factors, as well as inherited genetic factors, play in epigenetic reprogramming during development. Several studies have suggested that epigenetics is integral to phenotypic plasticity [67, 68], and one of the earliest observations demonstrates that increased maternal care in the 1st week of life alters DNA structure of the glucocorticoid receptor (GR) gene promoter in the hippocampus of the offspring [69] and also significantly affects offspring reproductive strategies. Differences in DNA methylation patterns between offspring of high and low care mothers emerged over the 1st week of life, persisted into adulthood and were associated with altered histone acetylation and transcription factor binding to the GR promoter. These DNA methylation-induced changes in gene expression were regulated through modifications of chromatin structure, suggesting a causal relationship between maternally induced, epigenetic modification of gene expression and phenotypic responses in offspring [69]. How changes to the epigenome occur remains unclear, although evidence suggests that early life nutrition may play a role [68]. Classic studies have shown that the degree of imprinting in the agouti mutant gene is affected by maternal diet, highlighting the possibility that early life nutrition could modify epigenetic regulation of gene expression in the developing fetus [68, 70].

Conclusion

There is now conclusive evidence that reproductive health is influenced by early life events and modified by postnatal environmental factors. Disorders of reproductive health are increasingly common in both developed and developing societies and are a major economic, psychological and social burden. The interaction and conflict between early life developmental and postnatal environments results in significant changes in the maturation and function of the reproductive endocrine axis, and the effects on future generations' reproductive potential is not known. Still, is it possible that over the last 100 years we have reset our trajectory for reproductive development in a way that may serve as an advantage? For example, it has been proposed that fetal growth restriction, as a consequence of impaired intrauterine conditions, one of which may be suboptimal nutrition, is one component of life history strategy where the organism predicts a shorter life and invests less into growth [71] but preserves ovarian function to maintain and even enhance reproductive fitness [72]. Indeed, even PCOS has been proposed to serve as advantageous rather than deleterious, given the right nutritional environment [73, 74]. Early life environmental effects are not isolated to the prenatal time period, and one must consider all early life windows – prenatal, lactational and prepubertal – within which energy resources must be partitioned between a number of biological systems (development, growth, reproduction) in a manner that best suits the environmental constraints to maximize the organism's fitness [75, 76]. It is within these systems therefore that trade-offs may appear, impairing one system in favour of another (growth versus reproduction, for example). Currently, the impact of global nutritional (high-fat diets) and chemical (toxins) environments on reproductive health and disease risk is poorly understood, but improved insight into underlying mechanisms is likely to have significant implications not only for current understanding of reproductive disorders, but also for future generations' reproductive potential.

References

1 Hanson MA, Gluckman PD: Developmental origins of health and disease: new insights. Basic Clin Pharmacol Toxicol 2008;2:90–93.

2 Gluckman PD, HansonMA: Changing times: the evolution of puberty. Mol Cell Endocrinol 2006;26–31.

3 Dumesic DA, Schramm RD, Abbott DH: Early origins of polycystic ovary syndrome. Reprod Fertil Dev 2005;17:349–360.

4 Tanner JM: Relation between age at puberty and adult physique in man. J Physiol 1955;1:17P.

5 Mills JL, Shiono PH, Shapiro LR, Crawford PB, Rhoads GG: Early growth predicts timing of puberty in boys: results of a 14-year nutrition and growth study. J Pediatr 1986;3:543–547.

6 Francois I, de Zegher F: Adrenarche and fetal growth. Pediatr Res 1997;3:440–442.

7 Ibanez L, Potau N, Francois I, de Zegher F: Precocious pubarche, hyperinsulinism, and ovarian hyperandrogenism in girls: relation to reduced fetal growth. J Clin Endocrinol Metab 1998;3558–3562.

8 Ibanez L, Ferrer A, Marcos MV, Hierro FR, de Zegher F: Early puberty: rapid progression and reduced final height in girls with low birth weight. Pediatrics 2000;5:e72.

9 Ibanez L, Dimartino-Nardi J, Potau N, Saenger P: Premature adrenarche – normal variant or forerunner of adult disease? Endocr Rev 2000;6:671–696.

10 Ibanez L, Potau N, Enriquez G, Marcos MV, de Zegher F: Hypergonadotrophinaemia with reduced uterine and ovarian size in women born small-for-gestational-age. Hum Reprod 2003;8:1565–1569.

11 Ibanez L, Ong K, Dunger DB, de Zegher F: Early development of adiposity and insulin resistance after catch-up weight gain in small-for-gestational-age children. J Clin Endocrinol Metab 2006;6:2153–2158.

12 Ong K, Potau N, Petry CJ, et al: Opposing influences of prenatal and postnatal weight gain on adrenarche in normal boys and girls. J Clin Endocrinol Metab 2004;6:2647–2651.

13 Ibanez L, Valls C, Potau N, Marcos MV, de Zegher F: Polycystic ovary syndrome after precocious pubarche: ontogeny of the low-birthweight effect. Clin Endocrinol 2001;5:667–672.

14 Kousta E: Premature adrenarche leads to polycystic ovary syndrome? Long-term consequences. Ann NY Acad Sci 2006;1:148–157.

15 Karlberg J: Secular trends in pubertal development. Horm Res 2002;19–30.

16 Kaltiala-Heino R, Kosunen E, Rimpela M: Pubertal timing, sexual behaviour and self-reported depression in middle adolescence. J Adolesc 2003;5:531–545.

17 Ibanez L, Potau N, Chacon P, Pascual C, Carrascosa A: Hyperinsulinaemia, dyslipaemia and cardiovascular risk in girls with a history of premature pubarche. Diabetologia 1998;9:1057–1063.

18 Frontini MG, Srinivasan SR, Berenson GS: Longitudinal changes in risk variables underlying metabolic syndrome X from childhood to young adulthood in female subjects with a history of early menarche: The Bogalusa Heart Study. Int J Obes Relat Metab Disord 2003;11:1398–1404.

19 Maskarinec G, Zhang Y, Takata Y, Pagano I, Shumay DM, Goodman MT, et al: Trends of breast cancer incidence and risk factor prevalence over 25 years. Breast Cancer Res Treat 2006;98:45–55.

20 Biro FM, Khoury P, Morrison JA: Influence of obesity on timing of puberty. Int J Androl 2006;1:272–277; discussion 86–90.

21 Cooper C, Kuh D, Egger P, et al: Childhood growth and age at menarche. Br J Obstet Gynaecol 1996;8:814–817.

22 Sloboda DM, Hart R, Doherty DA, Pennell CE, Hickey M: Age at menarche: influences of prenatal and postnatal growth. J Clin Endocrinol Metab 2007;92:46–50.

23 Ibanez l, de Zegher F: Puberty after prenatal growth restraint. Horm Res 2006;112–115.

24 Ibanez L, Potau N, Ferrer, Rodriguez-Hierro F, Marcos MV, de Zegher F: Reduced ovulation rate in adolescent girls born small for gestational age. J Clin Endocrinol Metab 2002;7:3391–3393.

25 Ibanez L, Valls C, Cols M, Ferrer A, Marcos MV, De Zegher F: Hypersecretion of FSH in infant boys and girls born small for gestational age. J Clin Endocrinol Metab 2002;5:1986–1988.

26 Cresswell JL, Egger P, Fall CH, Osmond C, Fraser RB, Barker DJ: Is the age of menopause determined in utero? Early Hum Dev 1997;143–148.

27 Ibanez L, Valls C, Ong K, Dunger DB, de Zegher F: Metformin therapy during puberty delays menarche, prolongs pubertal growth, and augments adult height: a randomized study in low-birthweight girls with early-normal onset of puberty. J Clin Endocrinol Metab 2006;91:2068–2073.

28 Ibanez L, Lopez-Bermejo A, Diaz M, Marcos MV, de Zegher F: Metformin treatment for 4 years to reduce total and visceral fat in low-birthweight girls with precocious pubarche. J Clin Endocrinol Metab 2008;93:1841–1845.

29 Parent AS, Teilmann G, Juul A, Skakkebaek NE, Toppari J, Bourguignon JP: The timing of normal puberty and the age limits of sexual precocity: variations around the world, secular trends, and changes after migration. Endocr Rev 2003;5:668–693.

30 Teilmann G, Pedersen CB, Skakkebaek NE, Jensen TK: Increased risk of precocious puberty in internationally adopted children in Denmark. Pediatrics 2006;2:e391–e399.
31 Proos LA, Hofvander Y, Tuvemo T: Menarcheal age and growth pattern of Indian girls adopted in Sweden. I. Menarcheal age. Acta Paediatr Scand 1991;8–9: 852–858.
32 Teilmann G, Boas M, Petersen JH, Main KM, Gormsen M, Damgaard K, Brocks V, Skakkebaek NE, Jensen TK: Early pituitary-gonadal activation before clinical signs of puberty in 5- to 8-year-old adopted girls: a study of 99 foreign adopted girls and 93 controls. J Clin Endocrinol Metab 2007;7:2538–2544.
33 Azziz R: PCOS: a diagnostic challenge. Reprod Biomed Online 2004;644–648.
34 Hart R, Hickey M, Franks S: Definitions, prevalence and symptoms of polycystic ovaries and polycystic ovary syndrome. Best Pract Res Clin Obstet Gynaecol 2004;5:671–683.
35 Ibanez L, de Zegher F, Potau N: Premature pubarche, ovarian hyperandrogenism, hyperinsulinism and the polycystic ovary syndrome: from a complex constellation to a simple sequence of prenatal onset. J Endocrinol Invest 1998;9:558–566.
36 Unsworth WP, Taylor JA, Robinson JE: Prenatal programming of reproductive neuroendocrine function. the effect of prenatal androgens on the development of estrogen positive feedback and ovarian cycles in the ewe. Biol Reprod 2005;3:619–627.
37 Abbott DH, Barnet DK, Bruns CM, Dumesic DA: Androgen excess fetal programming of female reproduction: a developmental aetiology for polycystic ovary syndrome? Hum Reprod Update 2005;4:357–374.
38 Hart R, Norman R: Polycystic ovarian syndrome – prognosis and outcomes. Best Pract Res Clin Obstet Gynaecol 2006;5:751–778.
39 Sir-Petermann T, Maliqueo M, Angel B, Lara HE, Perez-Bravo F, Recabarren SE: Maternal serum androgens in pregnant women with polycystic ovarian syndrome: possible implications in prenatal androgenization. Hum Reprod 2002;10:2573–2579.
40 Jensen RB, Vielwerth S, Larsen T, Greisen G, Veldhuis J, Juul A: Pituitary-gonadal function in adolescent males born appropriate or small for gestational age with or without intrauterine growth restriction. J Clin Endocrinol Metab 2007;4:1353–1357.
41 Jensen TK, Jorgensen N, Punab M, Haugen TB, Suominen J, Zilaitiene B, Horte A, Andersen AG, Carlsen E, et al: Association of in utero exposure to maternal smoking with reduced semen quality and testis size in adulthood: a cross-sectional study of 1,770 young men from the general population in five European countries. Am J Epidemiol 2004;1:49–58.
42 Rowe PJ, Comhaire FH, Hargreave TB, Mellows HJ: WHO manual for the standard investigation and diagnosis of the infertile couple. Cambridge, Cambridge University Press, 1993.
43 Chow V, Cheung AP: Male Infertility. J Reprod Med 2006;3:149–156.
44 Francois I, de Zegher F, Spiessens C, D'Hooghe T, Vanderschueren D: Low birth weight and subsequent male subfertility. Pediatr Res 1997;6:899–901.
45 Cicognani A, Alessandroni R, Pasini A, Pirazzoli P, Cassio A, Barbieri E, Cacciari E: Low birth weight for gestational age and subsequent male gonadal function. J Pediatr 2002;3:376–379.
46 Garn SM: The secular trend in size and maturational timing and its implications for nutritional assessment. J Nutr 1987;5:817–823.
47 Adair LS: Size at birth predicts age at menarche. Pediatrics 2001;4:e59.
48 Parent AS, Rasier G, Gerard A, Heger S, Roth C, Mastronardi C, Jung H, Ojeda SR, Bourguignon JP: Early onset of puberty: tracking genetic and environmental factors. Horm Res 2005;41–47.
49 Frisch RE, Hegsted DM, Yoshinaga K: Body weight and food intake at early estrus of rats on a high-fat diet. Proc Natl Acad Sci U S A 1975;10:4172–4176.
50 Kirtley D, Maher R: Effect of an isocaloric high fat diet on initiation of puberty in osborne-mendel rats. Biol Reprod 1979;2:331–338.
51 Zambrano E, Rodriguez-Gonzalez GL, Guzman C, Garcia-Becerra R, Boeck L, Diaz L, Menjivar M, Larrea F, Nathanielsz PW: A maternal low protein diet during pregnancy and lactation in the rat impairs male reproductive development. J Physiol 2005;1:275–284.
52 Guzman C, Cabrera R, Cardenas M, Larrea F, Nathanielsz PW, Zambrano E: Protein restriction during fetal and neonatal development in the rat alters reproductive function and accelerates reproductive ageing in female progeny. J Physiol 2006;1:97–108.
53 Hilakivi-Clarke L, Clarke R, Onojafe I, Raygada M, Cho E, Lippman M: A maternal diet high in n-6 polyunsaturated fats alters mammary gland development, puberty onset, and breast cancer risk among female rat offspring. Proc Natl Acad Sci U S A 1997; 17:9372–9377.
54 Grandjean P, Bellinger D, Bergman A, Cordier S, et al: The Faroes statement: human health effects of developmental exposure to chemicals in our environment. Basic Clin Pharmacol Toxicol 2008;2:73–75.
55 Phillips KP, Foster WG: Key developments in endocrine disrupter research and human health. J Toxicol Environ Health B Crit Rev 2008;3–4:322–344.

56 Buck Louis GM, Gray LE Jr, Marcus M, Ojeda SR, et al: Environmental factors and puberty timing: expert panel research needs. Pediatrics 2008;(suppl 3):S192–S207.
57 Caserta D, Maranghi L, Mantovani A, Marci R, Maranghi F, Moscarini M: Impact of endocrine disruptor chemicals in gynaecology. Hum Reprod Update 2008;1:59–72.
58 Den Hond E, Schoeters G: Endocrine disrupters and human puberty. Int J Androl 2006;1:264–271.
59 Sumbayev VV, Jensen JK, Hansen JA, Andreasen PA: Novel modes of oestrogen receptor agonism and antagonism by hydroxylated and chlorinated biphenyls, revealed by conformation-specific peptide recognition patterns. Mol Cell Endocrinol 2008;287:30–39.
60 Blanck HM, Marcus M, Tolbert PE, Rubin C, Henderson AK, Hertzberg VS, Zhang RH, Cameron L: Age at menarche and tanner stage in girls exposed in utero and postnatally to polybrominated biphenyl. Epidemiology 2000;6:641–647.
61 Vasiliu O, Muttineni J, Karmaus W: In utero exposure to organochlorines and age at menarche. Hum Reprod 2004;7:1506–1512.
62 Gladen BC, Klebanoff MA, Hediger ML, Katz SH, Barr DB, Davis MD, Longnecker MP: Prenatal DDT exposure in relation to anthropometric and pubertal measures in adolescent males. Environ Health Perspect 2004;17:1761–1767.
63 Gladen BC, Ragan NB, Rogan WJ: Pubertal growth and development and prenatal and lactational exposure to polychlorinated biphenyls and dichlorodiphenyl dichloroethene. J Pediatr 2000;4:490–496.
64 Guo YL, Lambert GH, Hsu CC, Hsu MM: Yucheng: health effects of prenatal exposure to polychlorinated biphenyls and dibenzofurans. Int Arch Occup Environ Health 2004;3:153–158.
65 Yang CY, Yu ML, Guo HR, Lai TJ, Hsu CC, Lambert G, Guo YL: The endocrine and reproductive function of the female Yucheng adolescents prenatally exposed to PCBs/PCDFs. Chemosphere 2005;3:355–360.
66 Main KM, Kiviranta H, Virtanen HE, Sundqvist E, Tuomisto JT, Tuomisto J, Vartiainen T, Skakkebaek NE, Toppari J: Flame retardants in placenta and breast milk and cryptorchidism in newborn boys. Environ Health Perspect 2007;10:1519–1526.
67 Jablonka E, Lamb MJ: The changing concept of epigenetics. Ann N Y Acad Sci 2002;1:82–96.
68 Waterland RA, Michels KB: Epigenetic epidemiology of the developmental origins hypothesis. Annu Rev Nutr 2007;1:363–388.
69 Weaver IC, Cervoni N, Champagne FA, D'Alessio AC, Sharma S, Seckl JR, Dymov S, Szyf M, Meaney MJ: Epigenetic programming by maternal behavior. Nat Neurosci 2004;8:847–854.
70 Wolff GL, Kodell RL, Moore SR, Cooney CA: Maternal epigenetics and methyl supplements affect agouti gene expression in Avy/a mice. FASEB J 1998; 11:949–957.
71 Gluckman PD, Hanson MA: Evolution, development and timing of puberty. Trends Endocrinol Metab 2006;1:7–12.
72 Jasienska G, Thune I, Ellison PT: From the cover: fatness at birth predicts adult susceptibility to ovarian suppression: an empirical test of the predictive adaptive response hypothesis. Proc Natl Acad Sci U S A 2006;34:12759–12762.
73 Shaw LM, Elton S: Polycystic ovary syndrome: a transgenerational evolutionary adaptation. BJOG 2008;2:144–148.
74 Gleicher N, Barad D: An evolutionary concept of polycystic ovarian disease: does evolution favour reproductive success over survival? Reprod Biomed Online 2006;5:587–589.
75 Ellison PT: Energetics and reproductive effort. Am J Hum Biol 2003;3:342–351.
76 Gluckman PD, Hanson MA, Beedle AS: Early life events and their consequences for later disease: a life history and evolutionary perspective. Am J Hum Biol 2007;1:1–19.

Dr. Martha Hickey
School of Women's and Infants' Health
King Edward Memorial Hospital, University of Western Australia
Subiaco, Perth, WA 6009 (Australia)
Tel. +61 8 9340 1330, Fax +61 8 9381 3031, E-Mail MHickey@obsgyn.uwa.edu.au

Newnham JP, Ross MG (eds): Early Life Origins of Human Health and Disease.
Basel, Karger, 2009, pp 123–132

Developmental Origins of Immune Tolerance: Pathways and Influences

Susan L. Prescott

School of Paediatrics and Child Health, University of Western Australia, Perth, W.A., Australia

The epidemic rise of many autoimmune and allergic diseases (including food allergy, coeliac disease and type 1 diabetes) demonstrates that immune tolerance pathways are highly susceptible to environmental change [1]. Furthermore, the expression of many of these conditions in early life also suggests that 'failure' of immune tolerance is an early event in many individuals, emphasising the need to understand the developmental origins of these conditions.

This review explores evidence of pre-symptomatic defects in children who go on to develop disorders of tolerance (with a focus on allergic disease), as well as the evidence that these pathways can be modified by interventions to promote tolerance and prevent disease. The development of oral tolerance is an antigen-driven process that appears to logically depend on regular exposure to foods and other antigens during critical early stages of development. Successful tolerance process is also likely to depend on other conducive exposures (such as favourable gut colonisation, breast milk and/or other immunomodulatory factors). It is now increasingly evident that allergen-exposure is not the primary cause of the allergy epidemic, and that allergen avoidance may be unsuccessful, or even detrimental in allergy prevention. Indeed, rising rates of immune disease are likely to reflect a combination of many environmental changes which compromise tolerance. Understanding other early host-environment interactions in early life is essential to designing better strategies to prevent disease.

Immune Pathways Involved in the Development of Tolerance

In the context of allergy, tolerance pathways are still incompletely understood, and different processes are likely to determine tolerance to oral versus inhaled allergens. While similar cell types are likely to be involved, differences in antigen 'dose', route of encounter and concurrent exposures are likely to explain the different mechanisms of 'oral tolerance' (anergy or deletion of food-responsive T cell clones) versus 'inhalant tolerance' (determined by patterns of cytokine production by inhalant-responsive T cell clones). In broad terms, the immune system must tread a narrow path between adequate, defensive responses and inappropriate (pathological) immune responses (as in allergic disease and autoimmunity). This 'immune homeostasis' is achieved by a diverse group of cells with important regulatory functions (including but not limited to CD4+CD25+T regulatory cells, CD8+ T cells, epithelial cells, dendritic cells and other antigen-presenting cells, APCs). These cells are involved in programming and/or regulating effector T cells to determine either tolerance or reactivity to antigens. In early life, infants appear to depend on 'signals' from the microbial environment to mature both T helper type 1 (Th1) and regulatory immune function. These signals, mediated through pattern recognition receptors such as toll-like receptors (TLRs), appear essential to achieving the immunological balance required for (a) pathogen protection and (b) normal immune tolerance. This emphasises the importance of early microbial exposure in the development of immune tolerance (see below).

Evidence that Early Defects in Immune Pathways Are Associated with Subsequent Allergic Disease

There is accumulating evidence of early pre-symptomatic differences between atopic and non-atopic individuals at birth, although there is still a great deal of uncertainty surrounding the early immune events that lead to allergy.

Pre-Symptomatic Differences in Effector T Cells

One of the most consistently observed associations with allergic disease [2, and many others] is a relative impairment of neonatal T cell IFNγ production. Recently, we noted that IFNγ production is significantly correlated with PKCζ expression, and that PKCζ had a 'protective' relationship with allergic disease and sensitisation [3]. Further studies are underway to examine the significance of this finding, and the potential role in predicting allergy for more targeted prevention.

Pre-Symptomatic Differences in Regulatory T Cells?

Regulatory T cells (Tregs) are currently one of the most topical candidates in the aetiology of allergic disease. It is proposed that allergic disease may result from a relative imbalance between inappropriate effector T cell responses and inadequate suppressive activity by regulatory populations (including but not limited to Tregs). Despite this, there is still very little information on the development of Treg function in early life. One previous study has shown that neonates at 'high risk' of allergy have reduced capacity to generate putative T regulatory cell populations after lipopolysaccharide stimulation [4]. More recently, we provided further preliminary evidence that allergic disease may be associated with altered neonatal CD4+CD25+CD127$^{lo/-}$ Treg function. It remains possible that reduced ability to regulate effector T cell responses could be a logical contributing factor in the failure of normal tolerance mechanisms. Further studies are needed to determine the significance of these observations.

Pre-Symptomatic Differences in Antigen-Presenting Cells and Innate Immune Pathways?

There is currently no conclusive evidence that high risk infants have more immature APC IL-12 signalling to account for relatively weaker Th1 responses noted in these children (above). Although one study [5] suggested that IL-12 responses to allergens are lower in neonates with a family history of allergic disease, other studies have shown no differences in capacity of high risk neonates to mount IL-12 responses to bacterial lipopolysaccharide [6].

APCs, together with Tregs, depend heavily on TLR-mediated signalling for activation and maturation. We recently examined for differences in early TLR function in allergic infants, and found that infant allergic disease is associated with pre-symptomatic differences in TLR function, with more exaggerated hyper-responsiveness to TLR ligands. This was seen without corresponding maturation of effector T cell function (as assessed by PHA responses) [7]. These differences were apparent at birth, prior to colonisation, and imply either genetic and/or modulating environmental influences in utero. Assessment of postnatal responses is now underway to determine how TLR responses are modified during early colonisation and environment-driven immune maturation.

Genetic Factors That Contribute to Successful Tolerance

The success of tolerance is likely to depend on a wide range of predisposing genetic factors, and there are currently very little data on this. While potential genetic polymorphisms are of great importance in understanding the complex gene-environmental interactions during this period, genomic programming cannot account for the rising rates of disease. There has been a consequential shift in interest to epigenetic influences. The 'epigenome' represents an additional layer of information coded within the genome in a sequence-independent manner, which can be modified by environmental exposures. Specifically, environment-driven changes in the patterns of DNA methylation in proximal gene promoters, chromatin remodelling, histone modifications and/or non-coding RNA interactions can alter the patterns of gene expression and clinical phenotype. Moreover, these modifications can also be stably heritable without altering the underlying genomic sequence. This field has revolutionised understanding of the mechanistic basis for gene-environment interactions. While this is likely to be of central importance in the rising rates of disease, this area is still poorly understood in the context of allergy pathogenesis. The effects of candidate environmental factors for the allergy epidemic on methylation patterns (or other epigenetic factors) have not yet been investigated.

Evidence that Early Immune Pathways Can Be Modified by Environmental Exposures

Early environmental exposures can modify neonatal T cell function, although the mechanisms are not clear. Preliminary studies also suggest that even in utero exposures such as infection [8], maternal diet [9, 10] and smoking [11, 10] can influence early immune development.

Role of Microbial Exposures

Microbial exposure has a critical early role in promoting maturation of pathways that suppress allergic responses (APC, Treg and Th1 responses). There is evidence from animal models that bacterial endotoxin [12] or probiotic [13] administration in utero can prevent allergic inflammation. We have observed that neonates born in an environment with high microbial burden (Papua New Guinea) had significantly lower inflammatory cytokine responses and higher IL-10 response to TLR ligands compared with newborns born to Australian women with high socio-economic status (although other factors could also contribute to this) [14]. In the early postnatal period, there is some evidence that microbial burden downregulates TLR function. Although TLRs are highly expressed on the intestinal surface in the neonatal period, expression and function of these receptors are rapidly downregulated during normal colonisation [15, 16], protecting the neonate from excessive inflammatory responses to colonising commensal flora. There is also evidence that systemic TLR4-mediated responses are downregulated in the postnatal period, with significantly lower peripheral blood mononuclear cell responses (notably MyD88-dependent TNFα, IL-10 and IL-6 responses to endotoxin) by 2 months of age compared with birth [17]. Although the mechanism is not clear, this change in function coincides with the period of rapid colonisation. In later childhood, there is also some evidence that chronic exposure to high levels of bacterial endotoxin (in farmers' children) is associated with less allergic disease, and with downregulated TLR-mediated (IFNγ, IL-10, TNFα and IL-12) cytokine responses [18]. In this context, it may be expected that hyper-responsiveness may reflect reduced rather than increased chronic microbial exposure. Thus, variations in the *timing* and *duration* of microbial exposure (and associated effects on

TLR function) may provide the key to understanding apparent inconsistencies in the 'hygiene hypothesis'.

The Role of Allergen Exposure in Promoting Tolerance

Normal development of oral tolerance is an antigen-driven process, which may depend on regular exposure to foods and other antigens during a critical early window. Other events and exposures during first exposure to these proteins may critically affect this process. The establishment of healthy gut colonisation has been shown to be one essential factor in promoting tolerance to both allergens and self-antigens [1, 19, 20]. Delays in either colonisation [19] or antigen/allergen exposure [21, 22] can both lead to failure of oral tolerance. Conversely, allergen exposure too early when the gut colonisation and local immune networks are less established may increase the risk of allergic or autoimmune disease. Other immunomodulatory exposures may also play a key role in this period. For example, some studies also suggest that continued breastfeeding during introduction of complementary foods may also promote tolerance [23]. Maternal milk contains immune mediators which have been recently shown to be critical for tolerance development in animal models (see below) [24]. While avoiding complementary foods in the first 4 months appears to reduce the risk of early allergy [25–30], there is little evidence that avoidance beyond 4 months is beneficial [31]. Moreover, avoidance beyond 6 months has been associated with increased risk of allergic disease (food allergy, eczema, asthma) [21, 31, 32]. A number of recent studies suggest that exposure to specific foods in the 4- to 6-month age range may reduce the risk of food allergies [21] and autoimmunity [22, 33] compared with children first exposed either before or after this 'window'. Based on these and other observations, opinion leaders in Europe (ESPGHAN) [34], the USA [35] and Australia [36] have independently challenged the practice of delayed complementary feeding until after 6 months, and revised their guidelines for complementary feeding from between 4–6 months for children born in allergy-prone industrialised countries [34, 35, 37].

Role of Immunomodulatory Dietary Fatty Acids

In the 1990s, the initial studies to link these dietary n–3 polyunsaturated fatty acids to the rise in allergic diseases noted protective relationships between the consumption of oily fish in childhood and the development of asthma and wheeze [38, 39]. More recently, several more observational studies have reported reduced risk of childhood eczema [40] and asthma [41, 42] in association with high maternal fish oil consumption. A much larger UK cohort (n = 1,238) also suggested a link between cord blood n–3:n–6 levels and higher risk for eczema and late-onset wheeze; however, these associations were not significant after adjustment for multiple comparisons [43]. There is also a strong immunological basis for investigating fish oil in this context, including the differential effects of n–3 and n–6 polyunsaturated fatty acids on a number of aspects of cellular function including APC function [44, 45], T cell function and the production of inflammatory prostaglandins and leukotrienes [reviewed in 46]. The first large intervention using fish oil to reduce allergy risk did not commence supplementation until around 6 months of age and did not find any reduction in the development of atopy, asthma or other allergic disease by 5 years of age [47]. It has been postulated that earlier interventions may be more effective, and this is supported by observed effects of maternal fish oil supplementation on neonatal immune function [3, 9]. Currently, there are several larger studies in progress (in Europe and Australia) that are specifically designed to assess this, and the results are awaited with great interest.

Role of Breastfeeding

Breast milk contains many bioactive and immunomodulatory factors (including cytokines, antibodies, bacteria and fatty acids). Many of these could influence immune function. Animal studies have shown that maternal milk is more tolerogenic than formulae and that addition of cytokine TGFβ to formulae can increase its tolerogenicity. A recent landmark study showed that tolerance to allergens present in maternal milk was dependent on the presence of TGFβ (again in an animal model [24]). In humans, there is some evidence that continued breastfeeding during introduction of complementary foods is important for promoting tolerance [23]. There are inherent limitations in studies addressing the effects of breastfeeding on allergy risk, including: recruitment and reporting biases, perceptions modifying feeding practices, confounding factors, and the inability to randomise and blind. It remains logical that breastfeeding during the window of first complementary feeding would be ideal but more studies are needed.

Role of Pollutants (Smoking)

Potential influences of most environmental pollutants on the developing fetus are poorly understood, and tobacco smoke is the most well studied. This remains a common but avoidable toxic exposure, which has been clearly linked with decreased lung growth and subsequent asthma. It appears increasingly likely that tobacco smoke can also influence early immune function, with detectable alterations in cytokine production by the fetoplacental unit, as detected ex vivo in cord blood [48], as well as in patterns of fetal mononuclear cell responses in vitro [11] and TLR microbial responses [49]. Despite this however, the associations between early cigarette smoke exposure and subsequent allergic disease remain controversial. At present, the avoidance of smoking in pregnancy and the postnatal period remains the only clear and undisputed recommendation for the avoidance of asthma and respiratory disease.

Strategies to Promote Tolerance

The aim of intervention in early life is to promote favourable conditions for normal tolerance during critical periods of immune development. As indicated above, successful tolerance is likely to depend on other conducive exposures (such as favourable gut colonization [19], breast milk [23] and/or other nutritional immunomodulatory factors [9]). Thus, logical approaches can be considered broadly in 4 areas:

1 Promoting optimal colonisation and gut maturation.
2 Promoting optimal allergen exposure (timing, dose, interval and route).
3 Using immunomodulatory factors such as n–3 fatty acids and breast milk that may promote tolerogenic conditions during allergen encounter and processing.
4 Avoiding known toxins such as smoking that predispose to persistent disease.

At present, there are very few formal recommendations on these points as good evidence is still not available. Challenging many long-held concepts, there are now studies (in progress and in design) that will examine the hypothesis that earlier introduction and regular exposure to 'allergenic' foods (rather than avoidance) may reduce the risk of specific allergies to these foods. To promote early colonisation, there is preliminary evidence that supplementation with some probiotic strains may reduce atopic dermatitis (but not other allergy outcomes) [50, and now several others]. However, more studies are needed before recommendations can be made. Similarly, while exposure to fish oil in early life may have some beneficial effects, the role in allergy prevention is still unclear and large-scale trials are still in progress. While the role of breastfeeding in allergy prevention is unclear (with many contradicting studies), this is still recommended for other reasons, and there is some evidence that continued breastfeeding during introduction of complementary foods promotes tolerance [23].

Conclusion

Events during early immune development are critical for the establishment of normal tolerance. Understanding early host-environment interactions during this critical period is essential to addressing the rising rates of immune disease. This epidemic is likely to reflect the combined effects of many environmental changes that compromise tolerance. The multifactorial nature of these interactions has presented major challenges to research in this area. Traditional research approaches frequently fail to either examine or account for many of these interactions, and this may account for many of the conflicting findings in the literature and the slow progress in this area. This highlights the need for more integrated, multidisciplinary and multifactorial approaches. Finally, any strategies to prevent disease must be also considered in the context of these complex interactions.

References

1 Bach JF: The effect of infections on susceptibility to autoimmune and allergic diseases. N Engl J Med 2002;347:911–920.

2 Tang MLK, Kemp AS, Thorburn J, Hill D: Reduced interferon gamma secretion in neonates and subsequent atopy. Lancet 1994;344:983–985.

3 Prescott SL, Irvine J, Dunstan JA, Hii C, Ferrante A: Protein kinase-C zeta: a novel 'protective' neonatal T cell marker that can be up-regulated by allergy prevention strategies. J Allergy Clin Immunol 2007;120:200–206.

4 Haddeland U, Karstensen AB, Farkas L, Bo KO, Pirhonen J, Karlsson M, Kvavik W, Brandtzaeg P, Nakstad B: Putative regulatory T cells are impaired in cord blood from neonates with hereditary allergy risk. Pediatr Allergy Immunol 2005;16:104–112.

5 Gabrielsson S, Soderlund A, Nilsson C, Lilja G, Nordlund M, Troye-Blomberg M: Influence of atopic heredity on IL-4-, IL-12- and IFN-gamma-producing cells in in vitro activated cord blood mononuclear cells. Clin Exp Immunol 2001;126:390–396.

6 Prescott S, Taylor A, King B, Dunstan J, Upham J, Thornton C, Holt P: Neonatal IL-12 capacity is associated with variations in allergen specific immune responses in the neonatal and postnatal periods. Clin Exp Allergy 2003;33:566–572.

7 Prescott SL, Noakes P, Chow B, Breckler L, Thornton CA, Hollams EM, van den Biggelaar AHJ, Ali M, Tulic M: Presymptomatic differences in Toll-like receptor function in infants who develop allergy. J Allergy Clin Immunol 2008;122:391–399.

8 Matsuoka T, Matsubara T, Katayama K, Takeda K, Koga M, Furukawa S: Increase of cord blood cytokine-producing T cells in intrauterine infection. Pediatr Int 2001;43:453–457.

9 Dunstan J, Mori TA, Barden A, Beilin LJ, Taylor A, Holt PG, Prescott SL: Fish oil supplementation in pregnancy modifies neonatal allergen-specific immune responses and clinical outcomes in infants at high risk of atopy: a randomised controlled trial. J Allergy Clin Immunol 2003;112:1178–1184.

10 Devereux G, Barker RN, Seaton A: Antenatal determinants of neonatal immune responses to allergens. Clin Exp Allergy 2002;32:43–50.

11 Noakes PS, Holt PG, Prescott SL: Maternal smoking in pregnancy alters neonatal cytokine responses. Allergy 2003;58:1053–1058.

12 Blumer N, Herz U, Wegmann M, Renz H: Prenatal lipopolysaccharide-exposure prevents allergic sensitisation and airway inflammation, but not airway responsiveness in a murine model of experimental asthma. Clin Exp Allergy 2005;35:397–402.

13 Blumer N, Sel S, Virna S, Patrascan CC, Zimmermann S, Herz U, Renz H, Garn H: Perinatal maternal application of *Lactobacillus rhamnosus* GG suppresses allergic airway inflammation in mouse offspring. Clin Exp Allergy 2007;37:348–357.

14 van den Biggelaar AHJ, Roponen M, Nadal-Sims MA, Devitt C, Phuanukoonnon S, Tulic M, Siba P, Prescott SL, Holt PG: Neonatal innate BCG responses are population dependent: pro- versus anti-inflammatory responses in Australian and Papua New Guinean newborns, submitted.

15 Abreu MT, Vora P, Faure E, Thomas LS, Arnold ET, Arditi M: Decreased expression of Toll-like receptor-4 and MD-2 correlates with intestinal epithelial cell protection against dysregulated proinflammatory gene expression in response to bacterial lipopolysaccharide. J Immunol 2001;167:1609–1616.

16 Savidge TC, Newman PG, Pan WH, Weng MQ, Shi HN, McCormick BA, Quaroni A, Walker WA: Lipopolysaccharide-induced human enterocyte tolerance to cytokine-mediated interleukin-8 production may occur independently of TLR-4/MD-2 signaling. Pediatr Res 2006;59:89–95.

17 Yerkovich ST, Wikstrom ME, Suriyaarachchi D, Prescott SL, Upham JW, Holt PG: Postnatal development of monocyte cytokine responses to bacterial lipopolysaccharide. Pediatr Res 2007;62:1–5.

18 Braun-Fahrlander C, Riedler J, Herz U, Eder W, Waser M, Grize L, Maisch S, Carr D, Gerlach F, Bufe A, Lauener RP, Schierl R, Renz H, Nowak D, von Mutius E: Environmental exposure to endotoxin and its relation to asthma in school-age children. N Engl J Med 2002;347:869–877.
19 Sudo N, Sawamura S, Tanaka K, Aiba Y, Kubo C, Koga Y: The requirement of intestinal bacterial flora for the development of an IgE production system fully susceptible to oral tolerance induction. J Immunol 1997;159:1739–1745.
20 Wills-Karp M, Santeliz J, Karp CL: The germless theory of allergic disease: revisiting the hygiene hypothesis. Nat Rev Immunol 2001;1:69–75.
21 Poole JA, Barriga K, Leung DYM, Hoffman DR, Eisenbarth G, Rewers M, Norris JM: Timing of initial exposure to cereal grains and the risk of wheat allergy. Pediatrics 2006;117:2175–2182.
22 Norris JM, Barriga K, Hoffenberg EJ, Taki I, Miao D, Haas JE, Emery LM, Sokol RJ, Erlich HA, Eisenbarth GS, Rewers M: Risk of celiac disease autoimmunity and timing of gluten introduction in the diet of infants at increased risk of disease. JAMA 2005;293:2343–2351.
23 Ivarsson A, Hernell O, Stenlund H, Persson LA: Breast-feeding protects against celiac disease. Am J Clin Nutr 2002;75:914–921.
24 Verhasselt V, Milcent V, Cazareth J, Kanda A, Fleury S, Dombrowicz D, Glaichenhaus N, Julia V: Breast milk–mediated transfer of an antigen induces tolerance and protection from allergic asthma. Nat Med 2008;14:170–175.
25 Fergusson DM, Horwood LJ, Shannon FT: Risk factors in childhood eczema. J Epidemiol Community Health 1982;36:118–122.
26 Forsyth JS, Ogston SA, Clark A, Florey CD, Howie PW: Relation between early introduction of solid food to infants and their weight and illnesses during the first two years of life. BMJ 1993;306:1572–1576.
27 Kajosaari M, Saarinen UM: Prophylaxis of atopic disease by six months' total solid food elimination. Evaluation of 135 exclusively breast-fed infants of atopic families. Acta Paediatr Scand 1983;72:411–414.
28 Morgan J, Williams P, Norris F, Williams CM, Larkin M, Hampton S: Eczema and early solid feeding in preterm infants. Arch Dis Child 2004;89:309–314.
29 Wilson AC, Forsyth JS, Greene SA, Irvine L, Hau C, Howie PW: Relation of infant diet to childhood health: seven year follow up of cohort of children in Dundee infant feeding study. BMJ 1998;316:21–25.
30 Zutavern A, Brockow I, Schaaf B, Bolte G, von Berg A, Diez U, Borte M, Herbarth O, Wichmann HE, Heinrich J: Timing of solid food introduction in relation to atopic dermatitis and atopic sensitization: results from a prospective birth cohort study. Pediatrics 2006;117:401–411.
31 Filipiak B, Zutavern A, Koletzko S, von Berg A, Brockow I, Grubl A, Berdel D, Reinhardt D, Bauer CP, Wichmann HE, Heinrich J: Solid food introduction in relation to eczema: results from a four-year prospective birth cohort study. J Pediatr 2007;151: 352–358.
32 Zutevern A, von Mutius E, Harris J, Mills P, Moffat S, White C, Cullinan P: The introduction of solids in relation to asthma and eczema. Arch Dis Child 2004;89:303–308.
33 Norris JM, Barriga K, Klingensmith G, Hoffman M, Eisenbarth GS, Erlich HA, Rewers M: Timing of initial cereal exposure in infancy and risk of islet autoimmunity. JAMA 2003;290:1713–1720.
34 Agostoni C, Decsi T, Fewtrell M, Goulet O, Kolacek S, Koletzko B, Michaelsen KF, Moreno L, Puntis J, Rigo J, Shamir R, Szajewska H, Turck D, van Goudoever J: Complementary feeding: a commentary by the ESPGHAN Committee on Nutrition. J Pediatr Gastroenterol Nutr 2008;46:99–110.
35 Greer FR, Sicherer SH, Burks AW: Effects of early nutritional interventions on the development of atopic disease in infants and children: the role of maternal dietary restriction, breastfeeding, timing of introduction of complementary foods, and hydrolyzed formulas. Pediatrics 2008;121:183–191.
36 Prescott SL, Smith P, Tang MLK, Palmer DJ, Sinn J, Huntley SJ, Cormack BW, Susan, Heine RG, Gibson RA, Makrides M: The importance of early complementary feeding in the development of oral tolerance: concerns and controversies. Pediatr Allergy Immunol 2008;19:375–380.
37 Prescott SL, Tang ML: The Australasian Society of Clinical Immunology and Allergy position statement: summary of allergy prevention in children. Med J Aust 2005;182:464–467.
38 Peat J, Salome C, Woolcock A: Factors associated with bronchial hyper-responsiveness in Australian adults and children. Eur Respir J 1992;5:921–929.
39 Poysa L, Korppi M, Remes K, Juntunen-Backman K: Atopy in childhood and diet in infancy. A nine-year follow-up study. Allergy Proc 1991;12:107–111.
40 Oddy WH, de Klerk NH, Kendall GE, Mihrshahi S, Peat JK: Ratio of omega-6 to omega-3 fatty acids and childhood asthma. J Asthma 2004;41:319–326.

41 Sausenthaler S, Koletzko S, Schaaf B, Lehmann I, Borte M, Herbarth O, von Berg A, Wichmann HE, Heinrich J, for the LSG: Maternal diet during pregnancy in relation to eczema and allergic sensitization in the offspring at 2 y of age. Am J Clin Nutr 2007;85:530.
42 Salam MT, Li YF, Langholz B, Gilliland FD: Maternal fish consumption during pregnancy and risk of early childhood asthma. J Asthma 2005;42:513–518.
43 Newson RB, Shaheen SO, Henderson AJ, Emmett PM, Sherriff A, Calder PC: Umbilical cord and maternal blood red cell fatty acids and early childhood wheezing and eczema. J Allergy Clin Immunol 2004;114:531–537.
44 Weatherill AR, Lee JY, Zhao L, Lemay DG, Youn HS, Hwang DH: Saturated and polyunsaturated fatty acids reciprocally modulate dendritic cell functions mediated through TLR4. J Immunol 2005;174: 5390–5397.
45 Hughes DA, Pinder AC: Influence of n-3 polyunsaturated fatty acids (PUFA) on the antigen-presenting function of human monocytes. Biochem Soc Trans 1996;24:389S.
46 Calder PC: N-3 polyunsaturated fatty acids and inflammation: from molecular biology to the clinic. Lipids 2003;38:343–352.
47 Marks GB, Mihrshahi S, Kemp AS, Tovey ER, Webb K, Almqvist C, Ampon RD, Crisafulli D, Belousova EG, Mellis CM, Peat JK, Leeder SR: Prevention of asthma during the first 5 years of life: a randomized controlled trial. J Allergy Clin Immunol 2006;118: 53–61.
48 Macaubas C, de Klerk NH, Holt BJ, Wee C, Kendall G, Firth M, Sly PD, Holt PG: Association between antenatal cytokine production and the development of atopy and asthma at age 6 years. Lancet 2003;362: 1192–1197.
49 Noakes PS, Hale J, Thomas R, Lane C, Devadason SG, Prescott SL: Maternal smoking is associated with impaired neonatal toll-like-receptor-mediated immune responses. Eur Respir J 2006;28:721–729.
50 Kalliomäki M, Salminen S, Poussa T, Arvilommi H, Isolauri E: Probiotics and prevention of atopic disease: 4-year follow-up of a randomised placebo-controlled trial. Lancet 2003;361:1869–1871.

Prof. Susan L. Prescott
School of Paediatrics and Child Health
University of Western Australia, Princess Margaret Hospital
PO Box D184, Perth, WA 6001 (Australia)
Tel. +61 8 9340 8171, Fax +61 8 9388 2097, E-Mail sprescott@meddent.uwa.edu.au

Newnham JP, Ross MG (eds): Early Life Origins of Human Health and Disease.
Basel, Karger, 2009, pp 133–141

Developmental Programming of the Kidney

Mark J. Nijland · Peter W. Nathanielsz

Center for Pregnancy and Newborn Research, Department of Obstetrics and Gynecology, School of Medicine, University of Texas Health Science Center at San Antonio, San Antonio, Tex., USA

Developmental programming, defined as the response by the developing mammalian organism to a specific challenge during a critical time window that alters the trajectory of development qualitatively and/or quantitatively with resulting persistent effects on phenotype, is now recognized as an important determinant of adult health. Several forms of exposure during early development such as maternal nutrient deprivation, nutrient excess, exogenous glucocorticoid excess and endogenous glucocorticoid excess due to maternal stress, provide convincing supporting evidence. Since developing organisms pass more biological milestones before birth than at any other time in their lives, it is no surprise that significant alterations in the timing or nature of these developmental steps have consequences in terms of function in later life. Development of each individual's specific phenotype, although based on a specific genome, is significantly influenced by epigenetic/environmental factors. Thus, it is vitally important to understand early life gene-environment interactions that can predispose to adult disease.

Acceptance and understanding of developmental programming comes from human epidemiological studies and a wealth of carefully controlled animal investigations primarily in rodents and sheep. There are numerous reviews on the exposures, mechanisms and outcomes involved [1–5]. Since development of renal systems in precocial species is relatively complete in utero or early in postnatal life, it is important to focus on the fetal contribution to the etiology of chronic kidney disease. Support comes from observations that chronic kidney disease [6], end-stage renal disease [7, 8] and low glomerular filtration rate (GFR) [9] are associated with low birth weight.

Critical Periods of Vulnerability

The idea that there are critical time windows when developing systems are most vulnerable to challenge, and the fact that these vulnerable periods differ among systems and species, is clearly supported by observations in animals.

In humans, the pronephros is a rudimentary structure that appears during week 4 of development (10% of gestation). The mesonephros develops caudal to the degenerating pronephros beginning at the end of the same week and achieves a degree of functional capacity in that it becomes perfused. Ureteric bud formation from the mesonephric duct marks the beginning of metanephric development during week 5 (12.5%) and full functionality, at least in urine production, is achieved by week 9 (22.5%). The intricate processes of tubulogenesis, glomerulogenesis and vasculogenesis that comprise nephron formation accelerate through mid-gestation before slowing to completion in late gestation, with nephron complement complete between 34 and 36 weeks (85–90%) of gestation. Continued organ growth occurs after birth but nephron number remains unchanged. Nephrogenesis is also complete before birth in nonhuman primate species [10, 11].

The sheep is perhaps the most significant long-gestation animal in which renal development is well characterized that delivers precocial offspring [12]. The pronephros does not occur in sheep (gestation length of 150 days); rather, a large glomerulus forms at the anterior end of the mesonephros, becomes functionally capable of producing urine, and is present from 11 to 38% of gestation. Much like the human, metanephric development begins at 18% and is complete by 90% of gestation in sheep [13]. Another well-studied species producing precocial young is the guinea pig. Metanephric development in this species begins around day 23 (34%) and is also complete before birth [14].

The vast majority of studies on consequences of developmental programming for kidney development have been conducted in rodents, particularly the rat. The most significant differences between this model and the others discussed so far, as well as the human, include the polytocous nature of rat pregnancy, altricial state of newborn rat pups and postnatal completion of metanephric development. In the rat, early renal development seems to occur rapidly, although as a percent of gestation it is somewhat slower compared with other species. The pronephros appears on embryonic day 10 (48% of gestation) followed by the appearance of the mesonephros on the same day. By day 12, the mesonephros is at the height of its development and on day 13 (62% of gestation) ureteric bud formation occurs and metanephric kidney development starts. The first primitive Loop of Henle forms around day 17 (80%), with the last appearing approximately on postnatal day 15 [15].

The Timing of Compromise

Some of the best evidence for developmental programming in humans comes from the study of individuals who survived the Dutch Hunger Winter of 1944–1945. Children who passed any period of their prenatal development during that time suffered from poor nutrition. Epidemiological studies involving the Dutch Hunger Winter cohort, whose participants are now over sixty years old, have demonstrated that maternal undernutrition at different stages of pregnancy produced different outcomes. When individuals were studied in their early 50s [16], 12% of those exposed to famine in mid-gestation were found to have microalbuminuria (defined as albumin/creatinine ratio ≥2.5), compared with 7% of those who were not prenatally exposed to famine (odds ratio 2.1). Correcting for other conditions influencing microalbuminuria, such as blood pressure and diabetes, did not diminish the association (adjusted odds ratio 3.2). Importantly, the presence of microalbuminuria was independent of size at birth. As described above, mid-gestation is the time of rapid metanephric development in primates, suggesting that this period is indeed a critical window of human renal development.

The literature contains several studies conducted to discover associations between specific periods of development and renal growth and function. Using a 50% reduction in maternal dietary protein in rats, Langley-Evans et al. [17] demonstrated decreased kidney size relative to body size in offspring whether the deprivation occurred between days 0–7, days 8–14, days 15–22 or for the whole duration of pregnancy. The authors concluded '... *Fetal exposure to maternal low-protein diets for any period in gestation may program hypertension in the rat. Alterations to renal structure, renal hormone action or the hypothalamic-pituitary-adrenal axis may all play a role in the programming phenomenon, either independently or in concert* ...'. Since plasma renin activity was elevated in offspring exposed to low maternal protein intake in the first 2 weeks of gestation, and angiotensin II concentrations were low in those exposed only in the last week of gestation, this excellent study provides a clear demonstration of the complexities of critical windows of vulnerability during development.

Perspective

Despite the species differences in timing of birth, mammalian kidneys follow similar developmental pathways across species. Cell culture studies using glomeruli from rat, rabbit, dog, sheep and rhesus monkey kidneys show that the growth and morphologic features of each cell type follow similar patterns, including numbers of cells present and rates of division [18]. Gene expression in nephron cell culture studies is similar among several species when normalized to the time scale of renal morphogenesis for each species [19]. Granulated glomerular epithelial cells from 17 mammalian species demonstrate similarities in morphology between all species studied at an electron microscopic level [20]. Animal studies provide critical information that aids in understanding renal development across species, including primates such as humans, but they will differ widely in terms of their response to intragestational insults. Kidney development in the rat, for example, will be impacted by a variety of postnatal environmental scenarios that similar stages of renal development in precocial species will never face; ex utero nutrient and water availability and variance in maternal care and behavior are prime examples.

Permanent Effects on Structure and Function

By its definition, developmental programming results in qualitative and/or quantitative alteration in phenotype that leads to increased susceptibility to disease.

Size

A common characteristic in renal developmental programming is a decrease in nephron endowment. During nephrogenesis, undifferentiated mesenchymal precursor cells and nephron precursor cells undergo a sequential series of morphologically well-defined events comprised of condensation mesenchyme-to-epithelial transition to form the various segments of the nephron. Immature nephrons are found in regions that are close to the outer cortex, whereas more mature structures reside next to the medulla. Each stage of this complex process has been studied in detail, primarily in rats and mice, and the genes, protein products, and regulators of transcription and translation have been described in detail [for reviews, see 21].

A strong correlation between essential hypertension in adult life and reduced nephron number exists in humans [22, 23]. Decreased nephron number and increased glomerular volume is also associated with IUGR [24, 25], while decreased glomerular number correlates both with decreased birth weight and incidence of chronic kidney disease [26, 27]. The idea that compromised renal development, rather than a simple reduction in nephron number, contributes to chronic kidney disease is further substantiated by the fact the kidney donors undergoing uninephrectomy have minimal risk of developing renal or cardiovascular complications from the procedure [28]. Removal of one kidney on the 1st day of postnatal life in the rat results in hypertension in the adult [29], while uninephrectomy at 100 dGA in sheep results in compensatory nephrogenesis in utero with reduced GFR and increased blood pressure in 6-month-old female offspring [30]. Reductions in nephron endowment associated with decreased maternal nutrition during nephrogenesis have been described in sheep [31], rats [17, 32] and mice [33]. Decreased nephron endowment induced during active nephrogenesis is clearly what is important for later function, primarily because alterations in structure during this phase of organogenesis likely reflect organ-wide deficits at multiple levels.

Tissue Phenotype

Our work in sheep has shown that maternal global caloric restriction alters intrarenal immunoreactive AT_1, AT_2 and renin expression in gestational age and gender-specific ways [13]. We have also evaluated the effect of 30% global restriction from 30 to 90 days of gestation in the baboon on the intrarenal renin angiotensin system in male (M) and female (F) fetal kidneys at 90 days [50% gestation – unpubl. data]. Steady-state mRNA and protein were evaluated using the Human Genechip U133A 2.0 or Western blot in 6 ad libitum-fed controls (C; 3M, 3F) and 6 nutrient-restricted (NR; 3M, 3F) fetuses. AT_1 mRNA was increased (92%) in NR males (NRM) compared with C males (CM), C females (CF), and NR females (NRF). Both diet and the interaction between diet and gender were significant. There was no diet or gender effect on AT_2 mRNA expression. Renal AT_1 protein expression exhibited both diet and gender effects, with immunoreactivity being increased in NRM compared with all other groups. Conversely, AT_2 protein was decreased in NRM compared with CM, CF and NRF, resulting in an AT_1:AT_2 protein ratio that was increased in NRM compared with CM, CF and NRF. A diet × gender interaction was observed on ACE immunoreactivity. Finally, no group differences were found in maternal plasma cortisol concentrations at 90 days. We found that diet-induced alterations in mRNA and protein expression, and the AT_1:AT_2 protein ratio, occur in male fetuses rather than females and are independent of maternal cortisol levels, leading to the postulate that gender-based sensitivity to nutrient deficit likely reflects differences between trajectories of growth, development and perhaps caloric demand in male and female fetuses.

We have conducted transcriptome analysis of fetal renal gene expression in the same model of nonhuman primate maternal undernutrition as described above [34, 35]. The purpose of these studies was to determine significant changes in fetal renal genes and renal structure involved in key metabolic pathways that result from moderate global maternal nutrient restriction. Fetal kidneys were collected at cesarean section at 50% gestation during the period of active metanephric development in the baboon. Using the Affymetrix human gene array U133A Plus 2.0 'Lab on a Chip' system, gene expression profiling in kidneys of fetuses of NR mothers showed downregulation of genes in ontological pathways (http://www.geneontology.org) related to RNA, DNA, and protein biosynthesis, metabolism and catabolism, actin cytoskeleton assembly, and apoptosis compared with kidneys of fetuses from ad libitum-fed mothers. In contrast, genes involved in cell signal transduction, communication and transport pathways were upregulated. Histological analysis of fetal kidney sections showed decreased tubule cross-sections per unit area following maternal nutrient restriction. There was no apparent effect on glomerular number as quantified by number of glomerulus cross-sections per unit area. We proposed that these changes indicate that even the moderate level of maternal nutrient restriction imposed accelerates fetal renal differentiation and inhibits kidney structure development with the consequence of shortening critical phases of renal growth.

In a subsequent study in the same model and at the same time in gestation, we conducted kidney transcriptome analysis by placing gene expression profile data into biological context using information technology methods (KEGG pathway analysis), thereby relating the expression profile to biochemical pathways [35]. By analyzing and comparing gene expression in kidneys of fetuses of control and nutrient-restricted mothers using an end-of-pathway approach, we identified key biochemical pathways contributing to the renal phenotype and influenced by decreased nutrient availability. Among the pathways identified in this manner, mammalian target of rapamycin signaling, which is upstream of overall translation into protein and plays a role in nutrient sensing, was highlighted as a promising candidate for examining mechanisms by which maternal nutrient restriction impacts renal phenotype.

Renal Function

Renal failure in Aboriginal Australians is reaching epidemic proportions [36]. In 1999, the incidence was more than 1,000 per million, doubling every 3–4 years. Hoy et al. [37] positively correlated low birth weight with body mass index (BMI), blood pressure, and diabetes rates. In their 1999 study, urinary albumin/creatine ratio was negatively correlated with birth weight. In overt albuminuria, the odds ratio was 2.8 after adjusting for other factors, such that low birth weight contributed to 27% of the prevalence of the condition in this high-risk population. The authors speculated that impaired nephrogenesis caused by intrauterine malnutrition was a likely cause [36]. IUGR and low normal renal function have recently been associated in 20- to 30-year-old participants of the Nord Trøndelag Health Study (1995–1997) in Norway [38]. Although the degree of impaired renal function was small in these young adults, it was significant and more consistent in men than women. Analysis of patients with end-stage renal disease using the Medical Birth Registry of Norway demonstrates that low birth weight is more strongly associated with development of the disease during the first 14 years of life than after age 15 years [39].

Lucas et al. [40, 41] investigated renal function and morphometry in 3-month-old rat offspring exposed to 50% global caloric restriction either during the first half, second half or throughout gestation. Nephron number, GFR and renal plasma flow were significantly lower in all nutrient-restricted groups when compared with controls, with a dose effect of duration of deprivation. In confirmation, a particularly noteworthy study has more recently focused on renal function in offspring of sheep that were nutrient restricted during pregnancy [42]. Total caloric intake was restricted in ewes from early to mid-gestation and the offspring examined at 1 year of age. GFR was quantified using renal scintigraphy (technetium-99-diethylenetriaminepentaacetic acid) to produce a dynamic renogram for each kidney. Free access to a high plane of nutrition after weaning, combined with a sedentary lifestyle, resulted in the offspring of both control-fed and nutrient-restricted ewes becoming overweight. In the overweight animals, blood glucose, calcium and leptin concentrations were increased, while estimated blood volume and plasma volume were significantly lower, compared with lean controls. Against this hematologic and endocrinologic background, the time to peak intensity of the renogram tended to by longer (left kidney: 39 and 23%; right kidney: 37 and 42%) while the slope of the renogram was significantly blunted in the overweight offspring. Absolute GFR tended to be higher and when corrected to lean body mass it tended to be lower in the overweight groups. The obesogenic lifestyle led by these offspring was also associated with marked renal histopathology as indicated by cortical apoptosis, glomerulosclerosis and luminal protein casts [42]. Among the more intriguing components of this study was the observation that prenatal undernutrition during nephrogenesis in the sheep prevented all of the obesity-induced renal histopathological changes, but did not alter the systemic cardiovascular dysfunction. The authors suggested that, while juvenile weight gain is contraindicated in achieving renal health, prenatal undernutrition may offer a degree of protection against renal damage despite the presence of an increase in blood pressure, hypovolemia, hyperglycemia, hypercalcemia and hyperleptinemia.

Perspective

From the clinical perspective it is hoped that a better understanding of developmental programming will lead to better diagnostic, preventative and therapeutic measures. Details of precise structural deficiencies need to be obtained experimentally. For example, as described above, the total number of nephrons may not be changed but tubule length and function may nonetheless be impacted, suggesting that size is not the major feature in programmed dysfunction. The persistence of programmed effects is likely due to covalent modifications of the genome resulting from changes in promoter methylation and histone acetylation. These epigenetic phenomena are central to the induction of persistent and heritable changes in gene expression that occur without alteration of DNA sequence [43–45]. Reversal of these molecular changes may be possible and may improve loss of function in existing structures but if developmental plasticity is no longer present, it may not be possible to reverse the structural changes. However, it is difficult to see how effects of a significant deficit in nephron endowment can be made good.

Conclusions

Chronic kidney disease in children and young adults predisposes to increased morbidity and mortality. According to the National Kidney Foundation and the National Institutes of Health, chronic kidney disease decreases the lifespan of patients receiving dialysis treatment by 40–60 years and of renal transplant recipients by 20–25 years when compared with a race- and age-matched US population [46]. In adults, there are strong ties between chronic kidney disease and cardiovascular disease, left ventricular hypertrophy and hypertension. Given the evidence presented in the preceding sections, we propose that developmental programming of renal development is a key component contributing to these important public health issues.

References

1 Gilbert JS, Nijland MJ: Sex differences in the developmental origins of hypertension and cardiorenal disease. Am J Physiol Regul Integr Comp Physiol 2008;295:R1941–R1952.

2 Symonds ME: Integration of physiological and molecular mechanisms of the developmental origins of adult disease: new concepts and insights. Proc Nutr Soc 2007;66:442–450.

3 Gilbert JS, Cox LA, Mitchell G, Nijland MJ: Nutrient-restricted fetus and the cardio-renal connection in hypertensive offspring. Expert Rev Cardiovasc Ther 2006;4:227–237.

4 Myatt L: Placental adaptive responses and fetal programming. J Physiol 2006;572:25–30.

5 Maloney CA, Rees WD: Gene-nutrient interactions during fetal development. Reproduction 2005;130: 401–410.

6 Li S, Chen S-C, Shlipak M, et al: Low birth weight is associated with chronic kidney disease only in men. Kidney Int 2008;73:637–642.

7 Fine RN: Management of growth retardation in pediatric recipients of renal allografts. Nat Clin Pract Nephrol 2007;3:318–324.

8 Lackland DT, Bendall HE, Osmond C, Egan BM, Barker DJP: Low birth weights contribute to the high rates of early-onset chronic renal failure in the southeastern United States. Arch Intern Med 2000; 160:1472–1476.

9 Keijzer-Veen MG, Schrevel M, Finken MJ, et al: Microalbuminuria and lower glomerular filtration rate at young adult age in subjects born very premature and after intrauterine growth retardation. J Am Soc Nephrol 2005;16:2762–2768.

10 Gubhaju L, Black MJ: The baboon as a good model for studies of human kidney development. Pediatr Res 2005;58:505–509.

11 Matsell DG, Mok A, Tarantal AF: Altered primate glomerular development due to in utero urinary tract obstruction. Kidney Int 2002;61:1263–1269.

12 Wintour EM, Alcorn D, Albiston A, et al: The renin-angiotensin system and the development of the kidney and adrenal in sheep. Clin Exp Pharmacol Physiol Suppl 1998;25:S97–S100.

13 Gilbert JS, Ford SP, Lang AL, et al: Nutrient restriction impairs nephrogenesis in a gender-specific manner in the ovine fetus. Pediatr Res 2007;61:42–47.

14 Zoetis T, Hurtt ME: Species comparison of anatomical and functional renal development. Birth Defects Res B Dev Reprod Toxicol 2003;68:111–120.

15 Hebel R: Anatomy and Embryology of the Laboratory Rat. Wörthsee, BioMed, 1986.

16 Painter RC, Roseboom TJ, van Montfrans GA, et al: Microalbuminuria in adults after prenatal exposure to the Dutch Famine. J Am Soc Nephrol 2005;16:189–194.

17 Langley-Evans SC, Welham SJ, Jackson AA: Fetal exposure to a maternal low protein diet impairs nephrogenesis and promotes hypertension in the rat. Life Sci 1999;64:965–974.

18 Holdsworth SR, Glasgow EF, Atkins RC, Thomson NM: Cell characteristics of cultured glomeruli from different animal species. Nephron 1978;22:454–459.

19 Horster M: Nephron function and perinatal homeostasis. Annals Vet Res 1977;8:468–482.

20 Gall JA, Alcorn D, Butkus A, Coghlan JP, Ryan GB: Distribution of glomerular peripolar cells in different mammalian species. Cell Tissue Res 1986;244:203–208.

21 Ballermann BJ: Glomerular endothelial cell differentiation. Kidney Int 2005;67:1668–1671.

22 Brenner BM, Garcia DL, Anderson S: Glomeruli and blood pressure. Less of one, more the other? Am J Hypertens 1988;1:335–347.

23 Brenner BM, Chertow GM: Congenital oligonephropathy and the etiology of adult hypertension and progressive renal injury. Am J Kidney Dis 1994;23: 171–175.

24 Konje JC, Okaro CI, Bell SC, de Chazal R, Taylor DJ: A cross-sectional study of changes in fetal renal size with gestation in appropriate- and small-for-gestational-age fetuses. Ultrasound Obstet Gynecol 1997; 10:22–26.

25 Manalich R, Reyes L, Herrera M, Melendi C, Fundora I: Relationship between weight at birth and the number and size of renal glomeruli in humans: a histomorphometric study. Kidney Int 2000;58:770–773.

26 Hoy WE, Hughson MD, Singh GR, Douglas-Denton R, Bertram JF: Reduced nephron number and glomerulomegaly in Australian Aborigines: a group at high risk for renal disease and hypertension. Kidney Int 2006;70:104–110.

27 Hughson MD, Douglas-Denton R, Bertram JF, Hoy WE: Hypertension, glomerular number, and birth weight in African Americans and white subjects in the southeastern United States. Kidney Int 2006;69: 671–678.

28 Sansalone CV, Maione G, Aseni P, et al: Early and late residual renal function and surgical complications in living donors: a 15-year experience at a single institution. Transplant Proc 2006;38:994–995.

29 Woods LL, Weeks DA, Rasch R: Hypertension after neonatal uninephrectomy in rats precedes glomerular damage. Hypertension 2001;38:337–342.

30 Moritz KM, Wintour EM, Dodic M: Fetal uninephrectomy leads to postnatal hypertension and compromised renal function. Hypertension 2002;39:1071–1076.
31 Gilbert JS, Lang AL, Grant AR, Nijland MJ: Maternal nutrient restriction in sheep: hypertension and decreased nephron number in offspring at 9 months of age. J Physiol 2005;565:137–147.
32 Desai M, Gayle D, Babu J, Ross MG: Permanent reduction in heart and kidney organ growth in offspring of undernourished rat dams. Am J Obstet Gynecol 2005;193:1224–1232.
33 Hoppe CC, Evans RG, Bertram JF, Moritz KM: Effects of dietary protein restriction on nephron number in the mouse. Am J Physiol 2007;292:R1768–R1774.
34 Cox LA, Nijland MJ, Gilbert JS, et al: Effect of 30 per cent maternal nutrient restriction from 0.16 to 0.5 gestation on fetal baboon kidney gene expression. J Physiol 2006;572:67–85.
35 Nijland MJ, Schlabritz-Loutsevitch NE, Hubbard GB, Nathanielsz PW, Cox LA: Non-human primate fetal kidney transcriptome analysis indicates mammalian target of rapamycin (mTOR) is a central nutrient-responsive pathway. J Physiol 2007;579:643–656.
36 Hoy WE, Rees M, Kile E, Mathews JD, Wang Z: A new dimension to the Barker hypothesis: low birthweight and susceptibility to renal disease. Kidney Int 1999;56:1072–1077.
37 Hoy WE, Mathews JD, McCredie DA, et al: The multidimensional nature of renal disease: rates and associations of albuminuria in an Australian Aboriginal community. Kidney Int 1998;54:1296–1304.
38 Hallan S, Euser AM, Irgens LM, Finken MJJ, Holmen J, Dekker FW: Effect of intrauterine growth restriction on kidney function at young adult age: the Nord Trøndelag Health (HUNT 2) Study. Am J Kidney Dis 2008;51:10–20.
39 Vikse BE, Irgens LM, Leivestad T, Hallan S, Iversen BM: Low birth weight increases risk for end-stage renal disease. JASN 2008;19:151–157.
40 Lucas SR, Costa-Silva VL, Zaladek-Gil F: Effects of intrauterine undernutrition on the renal function of the progeny. Braz J Med Biol Res 1989;22:1303–1306.
41 Lucas SR, Zaladek-Gil F, Costa-Silva VL, Miraglia SM: Function and morphometric evaluation of intrauterine undernutrition on kidney development of the progeny. Braz J Med Biol Res 1991;24:967–970.
42 Williams PJ, Kurlak LO, Perkins AC, et al: Hypertension and impaired renal function accompany juvenile obesity: the effect of prenatal diet. Kidney Int 2007;72:279–289.
43 Akintola AD, Crislip ZL, Catania JM, et al: Promoter methylation is associated with the age-dependent loss of N-cadherin in the rat kidney. Am J Physiol 2008;294:F170–F176.
44 Turner BM: Cellular memory and the histone code. Cell 2002;111:285–291.
45 Wyrwoll CS, Mark PJ, Waddell BJ: Developmental programming of renal glucocorticoid sensitivity and the renin-angiotensin system. Hypertension 2007;50:579–584.
46 Weir MR, Dworkin LD: Antihypertensive drugs, dietary salt, and renal protection: how low should you go and with which therapy? Am J Kidney Dis 1998;32:1–22.

Dr. Mark J. Nijland
Department of Obstetrics and Gynecology
UTHSCSA School of Medicine
7703 Floyd Curl Drive, San Antonio, TX 78229 (USA)
Tel. +1 210 567 5055, Fax +1 210 567 5033, E-Mail nijland@uthscsa.edu

Newnham JP, Ross MG (eds): Early Life Origins of Human Health and Disease.
Basel, Karger, 2009, pp 142–163

Perinatal Appetite Programming

Mina Desai · Erin Keen-Rhinehart · Michael G. Ross

Department of Obstetrics and Gynecology, Harbor-UCLA Medical Center, Geffen School of Medicine at UCLA, Los Angeles, Calif., USA

Currently in the United States and Western societies, there are major health epidemics of obesity, hypertension, and diabetes [1]. In the United States, 66% of adults are overweight (BMI 25 to <30) and 33% are obese (BMI ≥30), representing a modern health crisis [2]. Perhaps of even greater concern is the dramatic increase in childhood obesity. As childhood obesity is a major risk factor for adult obesity [3], the 20% incidence of childhood obesity [4] portends a further increase in adult obesity. In conjunction with the rates of obesity, the incidence of hypertension and type II diabetes is increasing. The rapid increase in the prevalence of obesity has been partly attributed to environmental factors, including increased food availability, high fat, western diet, and reduced energy expenditures associated with changing work habits. However, recent studies suggest that gestational programming of the thrifty phenotype may be a predominant factor in the epidemic of obesity. The 'thrifty phenotype' hypothesis or 'programming' proposes that nutrition during development modifies obesity susceptibility by altering cellular, tissue and organ development and perturbing homeostatic regulatory mechanisms [5–7]. Our studies demonstrate developmental programming of offspring metabolic syndrome (obesity, hypertension, diabetes) as a result of in utero environmental alterations [8–10], and implicate enhanced appetite and adipogenesis as contributory factors [11, 12]. This chapter focuses on the role of perinatally programmed hyperphagia in adult obesity with specific emphasis on the role of anorexigenic hormones, leptin and insulin.

Intrauterine Growth Restriction and Programmed Obesity

Several maternal conditions have been linked to obesity in both human and rodent offspring. Paradoxically, both maternal caloric deprivation [13–16] and maternal obesity [17] produce obese offspring. Similarly, mothers with type 2 [18] diabetes tend to have obese progeny. Conversely, intrauterine growth-restricted (IUGR) newborns have an increased risk of adult obesity and metabolic syndrome. Although interest has often focused on the generational effect of maternal obesity and newborn macrosomia, gestational programming of IUGR has contributed importantly to the population shift towards obesity. Briefly (fig. 1), normal weight mothers most commonly gave birth to normal weight infants which developed into normal weight adults. There is a U shaped curve for the relationship between birth weight and metabolic disease such that those born small or large show increased risk of obesity later in life [19–21]. The current increased incidence of prematurity and IUGR has resulted in an increase in low birth weight offspring. When combined with improved neonatal survival and exposure to western diet later in life, this results in an increase in offspring programmed during development for adult hyperphagia and obesity. This process has accounted, to a significant degree, for the ongoing population shift toward an obese phenotype, with these second generation (obese) mothers ultimately giving birth to macrosomic newborns.

Epidemiologic Studies

In the late 1980s to 1990s, Barker et al. [22] provided epidemiologic evidence that small for gestational age and/or low birth weight human newborns have a paradoxically increased risk of obesity as adults. In studies of Hertfordshire men aged 65 years, metabolic syndrome was present in 30% of those born at ≤2.95 kg as compared to only 6% among those born at 4.31 kg or greater. Thus, the prevalence of metabolic syndrome fell progressively in both men and women, from those who had the lowest to those who had the highest birth weights [22, 23]. Recent epidemiologic studies further demonstrate that the low birth weight infants with rapid catch-up growth have higher risk of obesity, insulin resistance and hypertension [24, 25]. This increased propensity for adult obesity is also evident in formula-fed infants [26]. A striking 25–63% of adult diabetes, hypertension and coronary heart disease can be attributed to low birth weight with subsequent accelerated newborn-to-adolescent weight gain [5]. Additionally, several studies have addressed the association of small size at birth with measures of later central obesity [27–29] insulin resistance [30, 31], and the metabolic syndrome [25, 32]. Thus, an apparent paradox of increased adiposity at both ends of the birth weight spectrum exists (i.e. higher BMI with higher birth weight and increased central obesity with lower birth weight).

The classic Dutch Hunger Winter studies illustrated the importance of critical period of intrauterine exposure to famine. It was found that male offspring of mothers who had been food deprived during the first two trimesters of pregnancy had a higher incidence of obesity [15]. Conversely, the men exposed to famine in the last trimester did not exhibit increased risk of obesity. Likewise, the offspring of survivors of the siege of Leningrad were not obese [33, 34]. Some of these apparently conflicting findings may be explained by the timing of the respective perturbations in relationship to critical milestones in the development of the nervous system as well as the genetic

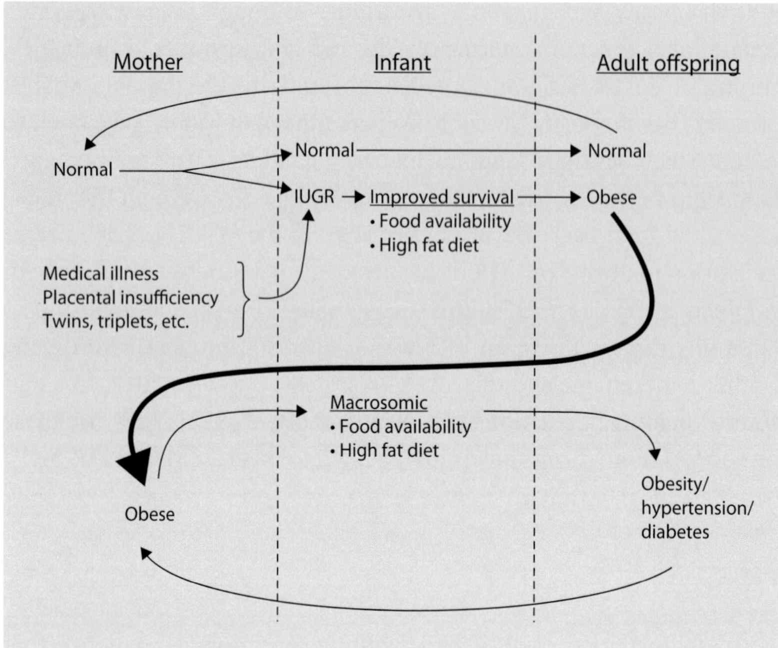

Fig. 1. Obese population shift due to programming.

background of the mother. It is thought that in early gestation, hypothalamic control of appetite becomes set in relation to body size [11]. Inappropriate setting of the hypothalamus could have led to obesity in men exposed to famine in utero.

Animal Studies

Prospective animal studies to test the 'fetal origins' or 'programming' hypothesis have imposed perturbations such as moderate to severe maternal undernutrition, feeding of low-protein diets throughout pregnancy, or glucocorticoid exposure during fetal development [35–38]. The large majority of studies confirm that perturbations of fetal growth and development lead to obesity [8, 14, 16, 39] in offspring. Furthermore, the prenatally food-restricted offspring are also hyperphagic [8, 10, 16, 40] and susceptible to stress-induced binge eating [41], suggesting altered appetite regulation as one of the underlying mechanism contributing to the development of obesity. Notably, high-fat diet during pregnancy [42, 43] and perinatal overnutrition [44–46] increases both the appetite and visceral adiposity of adult offspring. Thus, the common association of such diverse nutritional disturbances with hyperphagia and increased adiposity suggests that common mechanism(s) may be active in response to fetal nutrient imbalance. To discuss possible mechanisms, it is first essential to understand appetite regulation and development in an optimal nutritional environment.

Physiological Systems That Regulate Appetite

Hypothalamus

Although there are complex central mechanisms controlling energy homeostasis, the hypothalamus is recognized as the most critical appetite regulatory center. Appetite is primarily controlled by hypothalamic nuclei which receive input from central and peripheral sources including the brainstem, stomach (e.g. ghrelin), adipocytes (e.g. leptin) and pancreas (e.g. insulin; fig. 2). Together, these inputs provide neural and hormonal signals to balance the orexigenic drive to energetic needs. Within the hypothalamus, the arcuate nucleus (ARC) is a key target of appetite regulatory factors, and serves as a central neuronal processor of orexigenic and anorexigenic signals [47]. The ARC contains at least two distinct populations of neurons with opposing actions on food intake (fig. 2): primarily medial ARC orexigenic neurons containing neuropeptide Y (NPY) and agouti-related protein (AgRP) and lateral ARC anorexigenic neurons containing pro-opiomelanocortin (POMC) and cocaine- and amphetamine-regulated transcript (CART) [48]. ARC neurons also respond to blood-born signals (e.g. leptin, insulin and ghrelin) entering via capillaries of the median eminence, or cerebrospinal fluid signals diffusing from the third ventricle. ARC neurons exhibit high leptin, insulin and ghrelin receptor expression. Leptin and insulin are anorexigenic factors that stimulate the POMC/CART and inhibit the NPY/AgRP neurons, while ghrelin is an orexigenic factor that stimulates the NPY/AgRP and inhibits the POMC/CART neurons [48]. These hormones act at the ARC by affecting neuronal firing as well as transcription and translation of several genes related to energy homeostasis [47].

Many of the ARC NPY/AgRP and POMC/CART neurons project to downstream neurons in the paraventricular nucleus (PVN), which serves as an integrating center for appetite neural pathways [49]. POMC neurons mediate anorexigenic responses by release of α-melanocyte stimulating hormone (α-MSH) which binds with high affinity to PVN neurons expressing melanocortin 3 and 4 receptors (MC3/MC4-R) [50]. The orexigenic property of AgRP results from its competition with α-MSH at MC3/MC4Rs. Among the five subtypes of NPY receptors, NPY-1R, which is expressed on PVN neurons, is primarily responsible for NPY-induced increased food intake.

Additional Food Intake Pathways

The ARC interacts with additional hypothalamic nuclei including the ventromedial nucleus (VMH), lateral hypothalamus (LH) dorsomedial nucleus (DMH) and

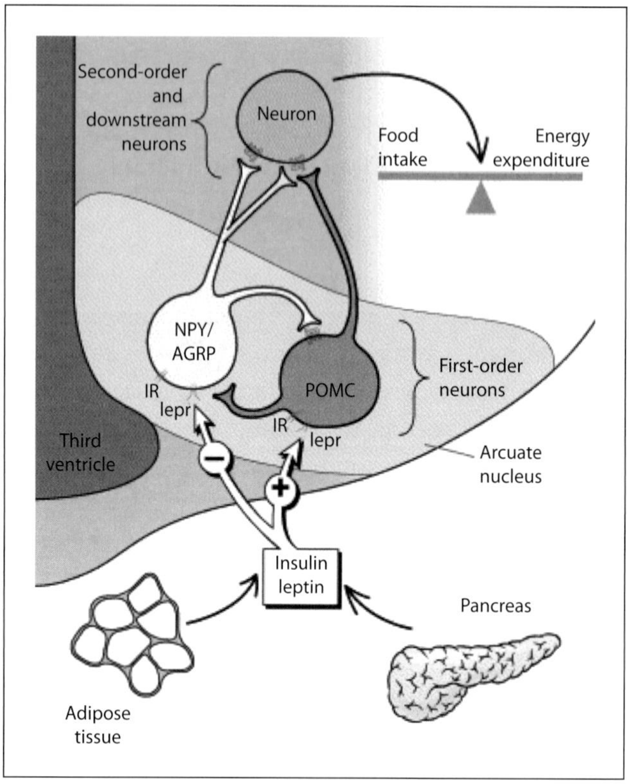

Fig. 2. Cell signaling in the ARC. From Niswender and Schwartz [160].

brainstem sites (e.g. locus coeruleus; nucleus of the solitary tract) [51, 52]. For example, the VMH, which contains neurons with a high level of leptin receptor expression, has been viewed as a 'satiety center', acting via connections with the PVN, LH and DMH. In contrast, the LH, which contains numerous neuronal subpopulations including those expressing the appetite stimulators melanin-concentrating hormone and the orexins, has been viewed as a 'feeding center'. The DMH, which contains both insulin and leptin receptors, communicates with the PVN, LH and brainstem regions. Outside of the hypothalamus, the medullary nucleus of the solitary tract integrates meal-related sensory information from the gastrointestinal tract. All of these other food-regulatory pathways play a significant role in appetite regulation that is most likely downstream or secondary to the ARC.

Anorexigenic Endocrine Factors

Appetite regulation is controlled by an interaction of diverse peripheral and central, endocrine and neuronal signals which influence short- and long-term orexigenic and anorexigenic responses [53].

Leptin
Leptin is the obesity (Ob) gene product, a 16-kDa protein synthesized primarily by adipocytes [54], and involved in hypothalamic appetite regulation. Leptin is released from adipocytes into the circulation and is transported across the blood brain barrier into the cerebrospinal fluid via a saturable transporter system in the choroid plexus, mediated by a short form of the leptin receptor (ObRa) [55]. Under most conditions studied, leptin induces satiety via upregulation of the ARC anorexigenic peptides (e.g. POMC/CART [56]) and downregulation of orexigenic peptides (e.g. NPY/AgRP). Thus, leptin levels are decreased by food restriction [57, 58] and increased by feeding, insulin, and glucose treatments. In sheep [59] and rodents [60], central leptin reduces voluntary food intake. Genetic mutations in the obese (ob) gene, which codes for leptin, or the diabetes (db) gene, which codes for leptin receptor, lead to hyperphagia and obesity [61, 62] due to unopposed NPY/AgRP signaling.

Leptin Signaling
In the ARC, both NPY/AgRP and POMC/CART neurons express cytokine receptor ObRb. Leptin binding to ObRb activates three main signaling cascades, JAK2-STAT3, MAPK, and PI3K, the latter two of which are also intermediates in insulin receptor (IR) activation [63, 64] (see below). ObRb binding recruits JAK2, which phosphorylates the receptor, after which the STAT3 monomer binds to the receptor and is subsequently phosphorylated (p-STAT3). Homodimerized p-STAT3 translocates to the nucleus and *activates* POMC transcription while *inhibiting* AgRP and NPY transcription. Leptin signaling through the JAK2-STAT3 pathway is under the negative feedback control of the endogenous inhibitor suppressor of cytokine signaling 3 (SOCS-3; fig. 3) as well as other factors (e.g. cytokine inhibitor signaling; protein tyrosine phophatase 1B) [65, 66].

Insulin
Like leptin, insulin affects hypothalamic neuropeptide expression. In the adult, little or no insulin is produced in the brain [67]. Thus, insulin enters brain regions in proportion with its plasma levels, acting in areas with dense IRs (e.g. ARC) [68, 69]. Similar to leptin, insulin gains access to the hypothalamus by means of a saturable receptor-mediated process and diffusion from the median eminence [70]. IRs are co-expressed on both POMC/CART and NPY/AgRP neurons.

Central insulin administration during rat food deprivation prevents the increase in ARC NPY mRNA and increases POMC mRNA [71]. The reduction in food intake caused by central insulin is blocked by a POMC antagonist [72]. Central insulin deficiency is a cause of hyperphagia; rats with untreated diabetes (insulin deficient), have reduced ARC POMC mRNA, increased NPY mRNA, and enhanced appetite, all of which are partially attenuated by peripheral insulin therapy [73], though the precise role of hyperinsulinemia on appetite in patients with type II diabetes has not been fully clarified. Of the four known mammalian isoforms of IR substrate (IRS), only IRS-2 is

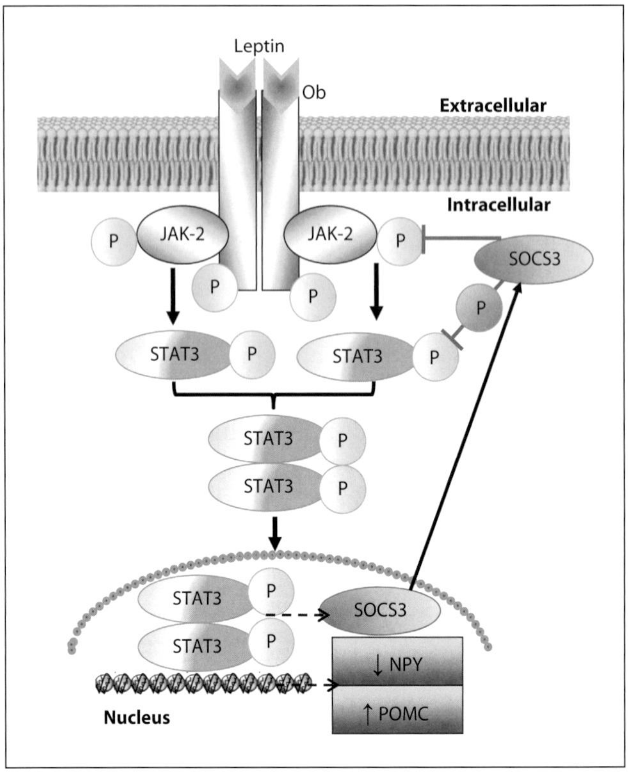

Fig. 3. Leptin signaling through the JAK2-STAT3 pathway.

implicated in neuronal control of energy homeostasis [74]. Thus, the IRS-2 knockout mouse exhibits an obese, hyperphagic phenotype. Hypothalamic insulin signaling also is required for inhibition of hepatic glucose production, indicating that central insulin resistance may contribute to hyperglycemia in type 2 diabetes mellitus [75].

Insulin Signaling

Insulin and leptin activate signaling via different classes of receptor molecules: leptin signals through a cytokine receptor as stated above, and insulin through a membrane-bound tyrosine kinase [76]. Following insulin binding, receptor autophosphorylation results in stimulation of the signaling cascade. Although the two main downstream pathways of insulin signaling involve activation of mitogen-activated protein kinases (MAPK; ERK1, ERK2) and PI3K, it is primarily the latter which regulates hypothalamic anorexigenic responses. IRS binds to phosphorylated residues on the IR and subsequently recruits the regulatory subunit p85 of PI3K. Activated PI3K phosphorylates PIP2 generating PIP3. The protein kinase B/AKT and PDK1 binds to PIP3 and phosphorylate AKT. p-AKT enters the nucleus where it phosphorylates FOXO1, which is subsequently excluded from the nucleus. In POMC neurons, FOXO1 diminishes POMC transcription by competing with binding sites for p-STAT3 in the promoter

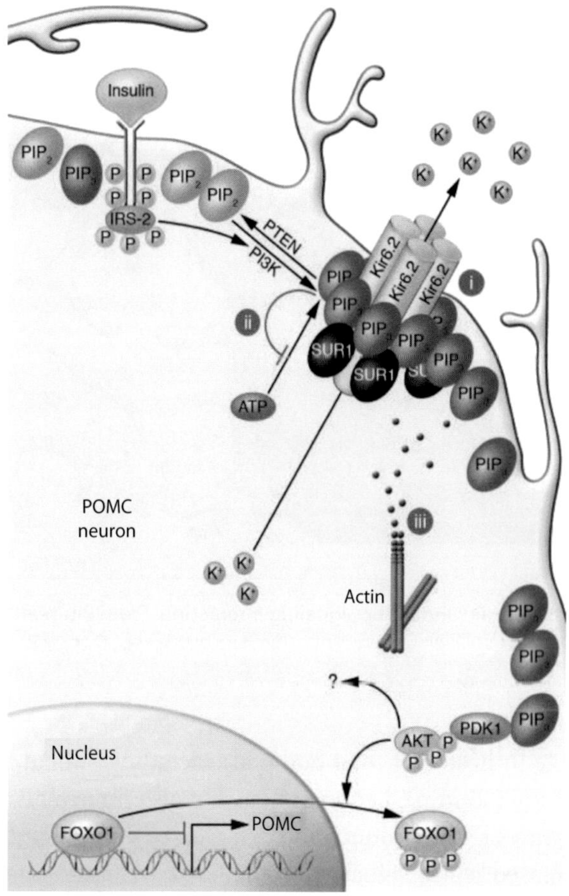

Fig. 4. Insulin PI3K signaling. From Plum et al. [76].

(fig. 4). Thus, insulin has the effect of increasing POMC transcription. In AgRP neurons, insulin also results in exclusion of FOXO1 from the nucleus. However, in contrast to POMC, FOXO1 increases AgRP transcription; thus, insulin acts to decrease AgRP transcription. Insulin signaling is inhibited by the lipid phosphatase PTEN which dephosphorylates PIP3 to PIP2.

Insulin and Leptin Signaling Interaction

Despite similarities between insulin and leptin, there are fundamental differences. Secretion of insulin is adjusted in response to acute changes in metabolism, increasing during meals or hyperglycemia and decreasing during exercise. The plasma half-life of insulin (2–3 min) is consistent with its acute regulatory role. Insulin secretion reflects visceral white adipose tissue, consistent with the elevated visceral fat in association with type 2 diabetes and metabolic syndrome. In contrast, leptin is a more

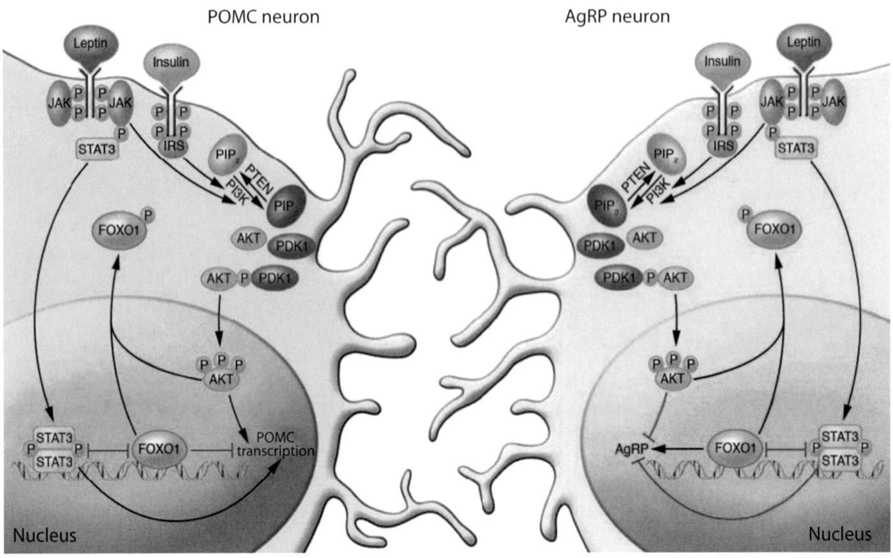

Fig. 5. Insulin and leptin signaling interaction. From Plum et al. [76].

stable indicator of total body fat (including subcutaneous fat) with a longer plasma half-life (45 min). Leptin and insulin also interregulate each other's secretion. Insulin increases leptin production via increased glucose utilization by adipocytes [77], while increased leptin signaling in the CNS increases whole-body insulin sensitivity [78]. When combined with signaling interactions (see below), these characteristics permit an integrated insulin/leptin anorexigenic regulatory system. Recently, leptin has been recognized as activating the PI3K signaling pathway [79]. Both insulin and leptin act to directly stimulate PI3K in POMC neurons, in the absence of synaptic transmission. In contrast, insulin and leptin show opposing effects in AgRP neurons [80] where leptin acts as an inhibitor (fig. 5). Of note, leptin-induced SOCS-3 also inhibits insulin signaling [81], though this has not yet been demonstrated in the hypothalamus.

Other Feeding-Related Signals

There is a diversity of appetite stimulatory and inhibitory factors. Ghrelin, synthesized in the stomach and hypothalamus [82, 83], is the endogenous ligand for the growth hormone secretagogue receptor [84, 85]. The ARC is the primary site of action [86, 87], where ghrelin increases expression of orexigenic and decreases expression of anorexigenic signals (fig. 1) [88]. Orexin A and B, peptides synthesized in the LH, are proposed to increase food intake partially via a Y1-receptor pathway [51]. Galanin, produced in several subpopulations of hypothalamic neurons, induces hyperphagia following ventricular or direct hypothalamic injections [51]. The PVN is innervated by both noradrenergic and dopaminergic fibers arising from the brain stem [89]. Noradrenergic receptors within the PVN and LH modulate feeding as satiated rats increase feeding in response to activation of adrenergic receptors within the PVN induced by either noradrenalin or clonidine [90, 91].

Insulin/Leptin as Neuronal Growth Factors

IRs are widely expressed throughout the developing brain, suggesting a role in neuronal growth [92, 93]. Although insulin has been considered a peripheral hormone, a series of studies [94, 95] have demonstrated the presence of insulin synthesis machinery in the fetal brain. Exogenous insulin promotes cell growth and serves as a trophic factor in fetal neuron cell culture [96]. Rat fetal brain cell culture at e16 expresses preproinsulin mRNA and insulin immunoreaction, and axonal growth is stimulated in insulin medium [97]. Insulin further stimulates MAPK phosphorylation, suggesting that neurite growth may be mediated via this signaling pathway, which has previously been implicated in neuronal development [98].

Leptin also has been recognized as a major neurotrophic factor during the development of the rodent hypothalamus, a critical neonatal period that precedes leptin's inhibition of food intake. Neural projection pathways from the ARC are permanently disrupted in leptin-deficient (ob/ob) mice and leptin treatment in adulthood does not reverse these neuroanatomical defects. However, treatment of ob/ob neonates with exogenous leptin rescues the development of ARC projections. Furthermore, leptin promotes neurite outgrowth from ARC neurons in vitro [99]. Neurosphere cultures in the presence of leptin maintained neural progenitors and increased their potential for astrocytic differentiation, implicating leptin in glial/neuronal development [100]. In addition, ob/ob mice have significantly higher levels of oligodendrocyte precursor cells than wild-type mice [101], confirming the role of leptin as a regulator of neural progenitor fate.

Development of Physiologic Systems Regulating Appetite

Although *fetal* growth is influenced by maternal nutritional, genetic and uteroplacental factors, orexigenic functions develop in utero to sustain newborn growth.

Rodents

In rodents, the hypothalamic appetite regulatory network is relatively immature at birth despite expression of appetite modulators. For instance, POMC becomes detectable in rat hypothalamic neurons from gestational day e12.5, whereas NPY neurons first appear in the ARC and dorsolateral hypothalamus at e14.5 [102]. Y1 receptor gene is expressed in rats beginning at e12 with specific binding occurring at e14 [103, 104]. Leptin signaling receptor immunoreactivity in the rat fetal PVN has also been reported [105].

Appetitive modulation of ingestive behavior becomes functional during the rat suckling newborn period, with maturation prior to the time of weaning [106]. More specifically, the pathway from the ARC to the DMH is mature by p6, and the pathway from the ARC to the PVN matures around p10. Some of the latest projections to develop form the ARC to LHA pathway, and they are complete between p12 and p14 [99, 107]. Notably, rats demonstrate a leptin surge at p10–12 which correlates with the development of ARC to PVN axonal projections [108]. There is continued increased plasma leptin from p12 through p20 with a decline on p21. In the mouse, ARC to PVN connections are established between p8 and p10 [109] and are dependent, in part, upon leptin trophic effects. Whereas leptin-mediated anorexigenic effects are mediated by the JAK/STAT pathway, trophic actions also may include the PI3K and MAP-ERK pathways [110, 111]. Thus, leptin and likely other factors contribute importantly to the development of hypothalamic feeding circuits. Neonatal overnourishment during this period inhibits ARC leptin responsiveness [112].

Humans

In humans, NPY immunoreactivity is detectable in 7-week embryos [113], with NPY immunopositive cells in the hippocampus at e105 and NPY nerve fibers at e112 [114]. NPY is present in the human hypothalamus as early as 21 weeks' gestation, and at this stage, there are already projections from the ARC to the PVN [115]. Immunoreactive ghrelin is also present in the umbilical cord as early as the 20th week of gestation [116]. Leptin is measurable in fetal plasma from e126 [117, 118]. In utero leptin levels are linked to fetal adiposity, as evidenced by the positive correlation between newborn serum leptin concentrations and birthweight [118]. Serum leptin levels in preterm newborns are lower than in term newborn infants [119], consistent with reduced preterm body fat. In infants, plasma leptin levels decrease with fasting and increase with feeding [120]. In contrast to rats, there is no evidence of a human postnatal leptin surge [121]. Unlike rodents, in humans the hypothalamic circuits regulating feeding develop primarily in utero during the last trimester of gestation, under direct influence of the maternal environment [115].

Newborn Orexigenic Drive

The normal development of orexigenic pathways results in enhanced orexigenic compared with anorexigenic function during early newborn life to facilitate growth. In newborn rats, NPY mRNA is observed in the ARC, the dorsomedial hypothalamic nucleus and the perifornical region [122], and as early as p2 are responsive to central NPY, demonstrating increased weight gain [123]. Although peripheral or central administration of leptin induces satiation in adults, leptin does not appear to acutely alter appetite in the fetus or newborn. In the near-term ovine fetus, central leptin does not inhibit swallowing of amniotic fluid or sucrose solutions [124, 125]. In near-term pig and rat fetuses, leptin receptors are weakly expressed within the hypothalamus as compared with the adult, and leptin receptor expression remains at a low level during the suckling period [126, 105]. In rat pups from p5 to p10, chronic leptin administration downregulates leptin receptor and NPY mRNA, but does *not* alter food intake [127]. Although mouse leptin levels increase 5- to 10-fold during the 2nd postnatal week, food intake during this period does not respond to treatment with exogenous leptin [128]. Thus, central leptin 'signaling responses' are developed in utero, though leptin does not regulate food intake during early development. The enhanced orexigenic vs. anorexigenic function likely serves to promote rapid newborn weight gain.

Mechanisms of Perinatal Appetite Programming

Hypothalamic Appetite-Regulating Network

Alterations in the hypothalamic appetite regulatory network clearly underlie the programmed hyperphagia that occurs in response to perinatal nutritional perturbations. These alterations bias the regulatory network toward hyperphagia by increasing production and sensitivity to orexigenic agents such as NPY. Among animal studies, ovine maternal food restriction results in near-term fetuses with increased hypothalamic NPY mRNA levels [129]. Similarly, in rats, maternal low-protein diet significantly increases NPY levels in the PVN and LHA of the offspring [130]. Utilizing brain slice electrophysiology techniques, it has been demonstrated that there is an enhancement of NPY inhibition of hypothalamic ARC anorexigenic neurons in adult offspring of maternal food-restricted rats during pregnancy [131]. Additional studies have shown that exposure to gestational and lactational maternal diabetes also causes similar effects in the hypothalamic appetite-regulating network with increased NPY and AgRP and decreased POMC and α-MSH [132, 133]. Finally, high-fat diet during gestation causes similar biasing toward hyperphagia by increasing offspring sensitivity to NPY and decreasing their sensitivity to leptin [134, 135].

Leptin Neuroendocrinology

Alterations in leptin physiology also appear to play a decisive role in programming hyperphagia in response to perinatal nutritional perturbations. In IUGR fetuses, placental leptin mRNA and protein and cord blood leptin levels are decreased [136, 137] and preterm or low birth weight human, rat, and calf newborns have reduced plasma leptin levels as well [138–140]. At 2 months of age, subcutaneous fat leptin mRNA is negatively correlated with birth weight [141]. As adults, leptin and insulin levels are inversely related to birth weight, independent of adult obesity [136]. The quality of nourishment during the neonatal period also impacts on leptin expression and function in later life, as preterm infants fed enriched formula had higher plasma leptin levels at 13–16 days than controls fed standard formula [142]. Childhood obesity predisposes to adult obesity in human [3, 143, 144] and subhuman primates [145].

Maternal 50% food restriction during the second half of rat gestation results in IUGR newborns with significantly decreased plasma leptin levels. When nursed by ad libitum-fed controls, IUGR offspring exhibit hyperphagia with adult obesity, marked by increased percent body fat and plasma leptin, suggesting altered anorexigenic pathways [8, 10]. Indeed, these offspring are resistant to the effects of anorexigenic and antiobesity agents such as leptin and sibutramine (a serotonin and noradrenalin reuptake inhibitor [11], indicating a permanent reduction in satiety signaling mechanisms. A similar phenomenon of hyperphagia, obesity and leptin resistance is observed in multiple species subjected to a nutritional deficit during gestation followed by nutritional excess shortly after birth [16, 146, 147].

Hypothalamic Anorexigenic Signaling Responses

We have undertaken a series of experiments to determine the underlying mechanism of programmed leptin resistance in IUGR offspring of gestationally food restricted dams. As early as 1 day of age, IUGR pups demonstrate dysregulation of central leptin and insulin signaling.

Basal

At 1 day of age, IUGR pups exhibit increased hypothalamic ObRb mRNA with elevated total STAT3 protein levels. At 3 weeks, these offspring show decreased hypothalamic ObRb mRNA expression though continued elevated STAT3 protein, suggesting a dysfunction in basal leptin signal [11]. Similarly, hypothalamic IR and insulin substrate (IRS-1 and IRS-2) mRNA expression is significantly reduced, suggesting that reduced insulin-mediated satiety responses of IUGR offspring may be responsible, in part, for hyperphagia observed in these offspring [148, 149].

Similar to models of gestational undernutrition, models of gestational overnutrition also cause alterations in leptin physiology. Maternal high-fat diet results in offspring that develop impaired leptin sensitivity [135] and impaired hypophagic response to insulin as adults [150, 151]. Interestingly, rats with genetic predisposition for diet-induced obesity have a preexisting reduction in central insulin sensitivity which is exacerbated by high-fat diet [152, 153].

Stimulated

To decipher the putative mechanism of programmed anorexigenic dysfunction in IUGR offspring, time sequence responses of hypothalamic JAK2-STAT3 signaling pathway to peripheral leptin have been studied. Preliminary results indicate that 1-day-old IUGR newborns have reduced p-STAT3 expression, suggesting sustained reduced leptin signaling, potentially a result of the failure to downregulate SOCS-3 [148, 149]. Further, as leptin and insulin signaling pathways are largely interdependent, the effects on insulin signaling molecules in response to peripheral leptin have also been studied. Remarkably, IUGR newborns exhibit a paradoxical increase in IRβ and IRS-2 proteins [148, 149]. Although these studies confirm an altered response to peripheral leptin, the contribution of altered peripheral to central transport (obRa) to leptin resistance is unknown. Together, however, these results indicate a marked dysfunction in basal and stimulated leptin and insulin signaling responses in the hypothalami from offspring of dams that were food restricted during pregnancy.

Thus, among IUGR newborns impaired anorexigenic mechanisms potentiate hyperphagia and rapid catch-up growth, ultimately resulting in increased offspring body fat, increased plasma leptin levels and a subsequent downregulation of leptin receptors (obRb) [154], consistent with leptin resistance [155, 154]. However, the increase in plasma leptin occurs beyond the critical window necessary for leptin induced feeding circuit development. Both developmentally altered neural pathways and leptin/insulin resistance at the level of the ARC likely contribute to adult hyperphagia.

Epigenetics

The demographic shift of populations toward a more obese phenotype in a relatively short period argues against a major genetic contribution and instead favors an environmental or epigenetic mechanism underlying this phenomenon. Alterations in DNA methylation and other epigenetic processes in multiple physiological systems have been demonstrated as a potential mechanism for developmental programming. For example, maternal food restriction that causes offspring hyperphagia also causes increased promoter methylation of energy balance-related receptors, namely the glucocorticoid receptor and the peroxisome proliferator-activated receptor in the liver [16, 156]. Further, hypermethylation in CpG islands of the functional SOCS-3 promoter is correlated with transcriptional silencing, which leads to constitutive activation of the JAK/STAT pathway in a cell culture model [157]. Thus, perinatal environment that impacts SOCS-3 methylation status may provide a potential mechanism for programmed obesity.

Critical Period of Leptin Treatment

Leptin acts primarily during a critical postnatal period to influence the development of hypothalamic appetite-regulatory network. In the rodent, this developmental window coincides with the natural surge in leptin. Treatment of leptin-deficient mice (ob/ob) with leptin does not restore ARC projections to the PVN, unless it occurs during p4–p12 [99]. Recent evidence also indicates that, in offspring of rat dams 70% nutrient deprived throughout gestation, leptin supplementation from p3 to p13 prevents programming of high susceptibility to diet-induced obesity [158]. Conversely, leptin supplementation to control pups from p1 to p10 causes adult hyperleptinemia, leptin resistance, increased food intake and excess body weight [154, 159]. Alternatively, limiting nutrient availability during the postnatal period by allowing food-restricted dams to nurse IUGR pups causes further newborn growth restriction, and prevents offspring hyperphagia and obesity [8].

Conclusions

Orexigenic function and regulation develop during in utero and early newborn periods. Programming of appetite/satiety mechanisms in response to an altered pregnancy/newborn environment influences infant, childhood and ultimately adult appetite 'setpoints'. In rodents, this period of plasticity occurs during the last week of gestation and the first 2 weeks of newborn life. In humans, the period of plasticity is less clearly defined but is likely to occur during the last trimester. Importantly, humans have well-developed orexigenic mechanisms, such that hunger and appetite are evoked (in response to nutrient restriction, hypoglycemia, etc.) initially with an unconscious drive and subsequently with increasing symptoms and conscious behavior changes. Conversely, satiety pathways are markedly less functional, contributing to the difficulty of self-motivated dietary weight loss. As demonstrated by our studies and others, gestational programming may further reduce the efficacy of satiety mechanisms, resulting in an appetite-satiety imbalance and hyperphagia-induced obesity. It is likely that IUGR programs impaired satiety responses by reduced anorexigenic neuronal development and altered cellular signaling pathways regulating both anorexigenic neuropeptide expression and neuronal activation.

References

1 James PT, Leach R, Kalamara E, Shayeghi M: The worldwide obesity epidemic. Obes Res 2001;9(suppl 4):228S–233S.

2 Ogden CL, Yanovski SZ, Carroll MD, Flegal KM: The epidemiology of obesity. Gastroenterology 2007;132:2087–2102.

3 Guo SS, Wu W, Chumlea WC, Roche AF: Predicting overweight and obesity in adulthood from body mass index values in childhood and adolescence. Am J Clin Nutr 2002;76:653–658.

4 Dietz WH: Health consequences of obesity in youth: childhood predictors of adult disease. Pediatrics 1998;101:518–525.

5 Barker DJ, Eriksson JG, Forsen T, Osmond C: Fetal origins of adult disease: strength of effects and biological basis. Int J Epidemiol 2002;31:1235–1239.

6 Forsen T, Osmond C, Eriksson JG, Barker DJ: Growth of girls who later develop coronary heart disease. Heart 2004;90:20–24.

7 Eriksson JG, Forsen T, Tuomilehto J, Osmond C, Barker DJ: Early adiposity rebound in childhood and risk of type 2 diabetes in adult life. Diabetologia 2003;46:190–194.

8 Desai M, Gayle D, Babu J, Ross MG: Programmed obesity in intrauterine growth-restricted newborns: modulation by newborn nutrition. Am J Physiol Regul Integr Comp Physiol 2005;288:R91–R96.

9 Desai M, Gayle D, Babu J, Ross MG: The timing of nutrient restriction during rat pregnancy/lactation alters metabolic syndrome phenotype. Am J Obstet Gynecol 2007;196:555–557.

10 Desai M, Babu J, Ross MG: Programmed metabolic syndrome: prenatal undernutrition and postweaning overnutrition. Am J Physiol Regul Integr Comp Physiol 2007;293:R2306–R2314.

11 Desai M, Gayle D, Han G, Ross MG: Programmed hyperphagia due to reduced anorexigenic mechanisms in intrauterine growth-restricted offspring. Reprod Sci 2007;14:329–337.

12 Desai M, Guang H, Ferelli M, Kallichanda N, Lane RH: Programmed upregulation of adipogenic transcription factors in intrauterine growth-restricted offspring. Reprod Sci 2008;15:785–796.

13 Barker M, Robinson S, Osmond C, Barker DJ: Birth weight and body fat distribution in adolescent girls. Arch Dis Child 1997;77:381–383.

14 Jones AP, Simson EL, Friedman MI: Gestational undernutrition and the development of obesity in rats. J Nutr 1984;114:1484–1492.

15 Ravelli GP, Stein ZA, Susser MW: Obesity in young men after famine exposure in utero and early infancy. N Engl J Med 1976;295:349–353.

16 Vickers MH, Breier BH, Cutfield WS, Hofman PL, Gluckman PD: Fetal origins of hyperphagia, obesity, and hypertension and postnatal amplification by hypercaloric nutrition. Am J Physiol Endocrinol Metab 2000;279:E83–E87.

17 Levin BE, Govek E: Gestational obesity accentuates obesity in obesity-prone progeny. Am J Physiol 1998;275:R1374–R1379.

18 Perseghin G, Ghosh S, Gerow K, Shulman GI: Metabolic defects in lean nondiabetic offspring of NIDDM parents: a cross-sectional study. Diabetes 1997;46:1001–1009.

19 Maffeis C, Micciolo R, Must A, Zaffanello M, Pinelli L: Parental and perinatal factors associated with childhood obesity in north-east Italy. Int J Obes Relat Metab Disord 1994;18:301–305.

20 Curhan GC, Willett WC, Rimm EB, Spiegelman D, Ascherio AL, Stampfer MJ: Birth weight and adult hypertension, diabetes mellitus, and obesity in US men. Circulation 1996;94:3246–3250.

21 Parsons TJ, Power C, Manor O: Fetal and early life growth and body mass index from birth to early adulthood in 1958 British cohort: longitudinal study. BMJ 2001;323:1331–1335.

22 Barker DJ, Hales CN, Fall CH, Osmond C, Phipps K, Clark PM: Type 2 (non-insulin-dependent) diabetes mellitus, hypertension and hyperlipidaemia (syndrome X): relation to reduced fetal growth. Diabetologia 1993;36:62–67.

23 Valsamakis G, Kanaka-Gantenbein C, Malamitsi-Puchner A, Mastorakos G: Causes of intrauterine growth restriction and the postnatal development of the metabolic syndrome. Ann N Y Acad Sci 2006; 1092:138–147.

24 Yajnik C: Interactions of perturbations in intrauterine growth and growth during childhood on the risk of adult-onset disease. Proc Nutr Soc 2000;59. 257–265.

25 Jaquet D, Deghmoun S, Chevenne D, Collin D, Czernichow P, Levy-Marchal C: Dynamic change in adiposity from fetal to postnatal life is involved in the metabolic syndrome associated with reduced fetal growth. Diabetologia 2005;48.849–855.

26 Bergmann KE, Bergmann RL, Von KR, Bohm O, Richter R, Dudenhausen JW, Wahn U: Early determinants of childhood overweight and adiposity in a birth cohort study: role of breast-feeding. Int J Obes Relat Metab Disord 2003;27:162–172.

27 Andrew R, Phillips DI, Walker BR: Obesity and gender influence cortisol secretion and metabolism in man. J Clin Endocrinol Metab 1998;83:1806–1809.

28 Ong KK, Ahmed ML, Emmett PM, Preece MA, Dunger DB: Association between postnatal catch-up growth and obesity in childhood: prospective cohort study. BMJ 2000;320:967–971.

29 Rogers I: The influence of birthweight and intrauterine environment on adiposity and fat distribution in later life. Int J Obes Relat Metab Disord 2003;27:755–777.

30 Yajnik C: Interactions of perturbations in intrauterine growth and growth during childhood on the risk of adult-onset disease. Proc Nutr Soc 2000;59:257–265.

31 Stocker CJ, Arch JR, Cawthorne MA: Fetal origins of insulin resistance and obesity. Proc Nutr Soc 2005;64:143–151.

32 Vanhala M: Childhood weight and metabolic syndrome in adults. Ann Med 1999;31:236–239.

33 Stanner SA, Yudkin JS: Fetal programming and the Leningrad Siege study. Twin Res 2001;4:287–292.
34 Stanner SA, Bulmer K, Andres C, Lantseva OE, Borodina V, Poteen VV, Yudkin JS: Does malnutrition in utero determine diabetes and coronary heart disease in adulthood? Results from the Leningrad siege study, a cross sectional study. BMJ 1997;315: 1342–1348.
35 McMillen IC, Robinson JS: Developmental origins of the metabolic syndrome: prediction, plasticity, and programming. Physiol Rev 2005;85:571–633.
36 Hales CN, Desai M, Ozanne SE: The Thrifty Phenotype hypothesis: how does it look after 5 years? Diabet Med 1997;14:189–195.
37 Desai M, Hales CN: Role of fetal and infant growth in programming metabolism in later life. Biol Rev Camb Philos Soc 1997;72:329–348.
38 Seckl JR, Meaney MJ: Glucocorticoid programming. Ann N Y Acad Sci 2004;1032:63–84.
39 Anguita RM, Sigulem DM, Sawaya AL: Intrauterine food restriction is associated with obesity in young rats. J Nutr 1993;123:1421–1428.
40 Vickers MH, Breier BH, McCarthy D, Gluckman PD: Sedentary behavior during postnatal life is determined by the prenatal environment and exacerbated by postnatal hypercaloric nutrition. Am J Physiol Regul Integr Comp Physiol 2003;285: R271–R273.
41 Hagan MM, Wauford PK, Chandler PC, Jarrett LA, Rybak RJ, Blackburn K: A new animal model of binge eating: key synergistic role of past caloric restriction and stress. Physiol Behav 2002;77:45–54.
42 Taylor PD, Poston L: Developmental programming of obesity in mammals. Exp Physiol 2007;92:287–298.
43 Beck B, Kozak R, Moar KM, Mercer JG: Hypothalamic orexigenic peptides are overexpressed in young Long-Evans rats after early life exposure to fat-rich diets. Biochem Biophys Res Commun 2006;342:452–458.
44 Plagemann A, Heidrich I, Gotz F, Rohde W, Dorner G: Obesity and enhanced diabetes and cardiovascular risk in adult rats due to early postnatal overfeeding. Exp Clin Endocrinol 1992;99:154–158.
45 Davidowa H, Plagemann A: Inhibition by insulin of hypothalamic VMN neurons in rats overweight due to postnatal overfeeding. Neuroreport 2001;12:3201–3204.
46 Davidowa H, Li Y, Plagemann A: Altered responses to orexigenic (AGRP, MCH) and anorexigenic (alpha-MSH, CART) neuropeptides of paraventricular hypothalamic neurons in early postnatally overfed rats. Eur J Neurosci 2003;18:613–621.
47 Barsh GS, Schwartz MW: Genetic approaches to studying energy balance: perception and integration. Nat Rev Genet 2002;3:589–600.
48 Meister B: Neurotransmitters in key neurons of the hypothalamus that regulate feeding behavior and body weight. Physiol Behav 2007;92:263–271.
49 Lambert PD, Phillips PJ, Wilding JP, Bloom SR, Herbert J: c-fos expression in the paraventricular nucleus of the hypothalamus following intracerebroventricular infusions of neuropeptide Y. Brain Res 1995;670:59–65.
50 Fan W, Dinulescu DM, Butler AA, Zhou J, Marks DL, Cone RD: The central melanocortin system can directly regulate serum insulin levels. Endocrinology 2000;141:3072–3079.
51 Kalra SP, Dube MG, Pu S, Xu B, Horvath TL, Kalra PS: Interacting appetite-regulating pathways in the hypothalamic regulation of body weight. Endocr Rev 1999;20:68–100.
52 Williams G, Bing C, Cai XJ, Harrold JA, King PJ, Liu XH: The hypothalamus and the control of energy homeostasis: different circuits, different purposes. Physiol Behav 2001;74:683–701.
53 Arora S, Anubhuti: Role of neuropeptides in appetite regulation and obesity – a review. Neuropeptides 2006;40:375–401.
54 Zhang Y, Proenca R, Maffei M, Barone M, Leopold L, Friedman JM: Positional cloning of the mouse obese gene and its human homologue. Nature 1994;372:425–432.
55 Jequier E: Leptin signaling, adiposity, and energy balance. Ann N Y Acad Sci 2002;967:379–388.
56 Korner J, Savontaus E, Chua SC Jr, Leibel RL, Wardlaw SL: Leptin regulation of Agrp and Npy mRNA in the rat hypothalamus. J Neuroendocrinol 2001;13:959–966.
57 Marie M, Findlay PA, Thomas L, Adam CL: Daily patterns of plasma leptin in sheep: effects of photoperiod and food intake. J Endocrinol 2001;170:277–286.
58 Dallongeville J, Hecquet B, Lebel P, Edme JL, Le Fur C, Fruchart JC, Auwerx J, Romon M: Short term response of circulating leptin to feeding and fasting in man: influence of circadian cycle. Int J Obes Relat Metab Disord 1998;22:728–733.
59 Morrison CD, Daniel JA, Holmberg BJ, Djiane J, Raver N, Gertler A, Keisler DH: Central infusion of leptin into well-fed and undernourished ewe lambs: effects on feed intake and serum concentrations of growth hormone and luteinizing hormone. J Endocrinol 2001;168:317–324.
60 Muzzin P, Cusin I, Charnay Y, Rohner-Jeanrenaud F: Single intracerebroventricular bolus injection of a recombinant adenovirus expressing leptin results in reduction of food intake and body weight in both lean and obese Zucker fa/fa rats. Regul Pept 2000;92: 57–64.

61 Clement K, Vaisse C, Lahlou N, Cabrol S, Pelloux V, Cassuto D, Gourmelen M, Dina C, Chambaz J, Lacorte JM, Basdevant A, Bougneres P, Lebouc Y, Froguel P, and Guy-Grand B: A mutation in the human leptin receptor gene causes obesity and pituitary dysfunction. Nature 1998;392:398–401.
62 Montague CT, Farooqi IS, Whitehead JP, Soos MA, Rau H, Wareham NJ, Sewter CP, Digby JE, Mohammed SN, Hurst JA, Cheetham CH, Earley AR, Barnett AH, Prins JB, O'Rahilly S: Congenital leptin deficiency is associated with severe early-onset obesity in humans. Nature 1997;387:903–908.
63 Maratos-Flier E: The long reach of leptin. Nat Med 2008;14:604–606.
64 Doherty GH, Oldreive C, Harvey J: Neuroprotective actions of leptin on central and peripheral neurons in vitro. Neuroscience 2008;154:1297–1307.
65 Cheng A, Uetani N, Simoncic PD, Chaubey VP, Lee-Loy A, McGlade CJ, Kennedy BP, Tremblay ML: Attenuation of leptin action and regulation of obesity by protein tyrosine phosphatase 1B. Dev Cell 2002;2:497–503.
66 Zabolotny JM, ce-Hanulec KK, Stricker-Krongrad A, Haj F, Wang Y, Minokoshi Y, Kim YB, Elmquist JK, Tartaglia LA, Kahn BB, Neel BG: PTP1B regulates leptin signal transduction in vivo. Dev Cell 2002;2:489–495.
67 Banks WA: The source of cerebral insulin. Eur J Pharmacol 2004;490:5–12.
68 Unger J, McNeill TH, Moxley RT III, White M, Moss A, Livingston JN: Distribution of insulin receptor-like immunoreactivity in the rat forebrain. Neuroscience 1989;31:143–157.
69 Wang J, Leibowitz KL: Central insulin inhibits hypothalamic galanin and neuropeptide Y gene expression and peptide release in intact rats. Brain Res 1997;777:231–236.
70 Schwartz MW, Figlewicz DP, Baskin DG, Woods SC, Porte D Jr: Insulin in the brain: a hormonal regulator of energy balance. Endocr Rev 1992;13:387–414.
71 Air EL, Benoit SC, Clegg DJ, Seeley RJ, Woods SC: Insulin and leptin combine additively to reduce food intake and body weight in rats. Endocrinology 2002;143:2449–2452.
72 Benoit SC, Air EL, Coolen LM, Strauss R, Jackman A, Clegg DJ, Seeley RJ, Woods SC: The catabolic action of insulin in the brain is mediated by melanocortins. J Neurosci 2002;22:9048–9052.
73 Sipols AJ, Baskin DG, and Schwartz MW: Effect of intracerebroventricular insulin infusion on diabetic hyperphagia and hypothalamic neuropeptide gene expression. Diabetes 1995;44:147–151.
74 Withers DJ: Insulin receptor substrate proteins and neuroendocrine function. Biochem Soc Trans 2001;29:525–529.
75 Obici S, Zhang BB, Karkanias G, Rossetti L: Hypothalamic insulin signaling is required for inhibition of glucose production. Nat Med 2002;8:1376–1382.
76 Plum L, Belgardt BF, Bruning JC: Central insulin action in energy and glucose homeostasis. J Clin Invest 2006;116:1761–1766.
77 Havel PJ: Role of adipose tissue in body-weight regulation: mechanisms regulating leptin production and energy balance. Proc Nutr Soc 2000;59:359–371.
78 van DG, de VK, Nyakas C, Buwalda B, Adage T, Kuipers F, Kas MJ, Adan RA, Wilkinson CW, Thiele TE, Scheurink AJ: Reduced anorexigenic efficacy of leptin, but not of the melanocortin receptor agonist melanotan-II, predicts diet-induced obesity in rats. Endocrinology 2005;146:5247–5256.
79 Niswender KD, Gallis B, Blevins JE, Corson MA, Schwartz MW, Baskin DG: Immunocytochemical detection of phosphatidylinositol 3-kinase activation by insulin and leptin. J Histochem Cytochem 2003;51:275–283.
80 Xu AW, Kaelin CB, Takeda K, Akira S, Schwartz MW, Barsh GS: PI3K integrates the action of insulin and leptin on hypothalamic neurons. J Clin Invest 2005;115:951–958.
81 Torisu T, Sato N, Yoshiga D, Kobayashi T, Yoshioka T, Mori H, Iida M, Yoshimura A: The dual function of hepatic SOCS3 in insulin resistance in vivo. Genes Cells 2007;12:143–154.
82 Kojima M, Hosoda H, Date Y, Nakazato M, Matsuo H, Kangawa K: Ghrelin is a growth-hormone-releasing acylated peptide from stomach. Nature 1999;402:656–660.
83 Cowley MA, Smith RG, Diano S, Tschop M, Pronchuk N, Grove KL, Strasburger CJ, Bidlingmaier M, Esterman M, Heiman ML, Garcia-Segura LM, Nillni EA, Mendez P, Low MJ, Sotonyi P, Friedman JM, Liu H, Pinto S, Colmers WF, Cone RD, Horvath TL: The distribution and mechanism of action of ghrelin in the CNS demonstrates a novel hypothalamic circuit regulating energy homeostasis. Neuron 2003;37:649–661.
84 Guan XM, Yu H, Palyha OC, McKee KK, Feighner SD, Sirinathsinghji DJ, Smith RG, Van der Ploeg LH, Howard AD: Distribution of mRNA encoding the growth hormone secretagogue receptor in brain and peripheral tissues. Brain Res Mol Brain Res 1997;48:23–29.
85 Katayama M, Nogami H, Nishiyama J, Kawase T, Kawamura K: Developmentally and regionally regulated expression of growth hormone secretagogue receptor mRNA in rat brain and pituitary gland. Neuroendocrinol 2000;72:333–340.

86 Cone RD, Cowley MA, Butler AA, Fan W, Marks DL, Low MJ: The arcuate nucleus as a conduit for diverse signals relevant to energy homeostasis. Int J Obes Relat Metab Disord 2001;25(suppl 5):S63–S67.
87 Hewson AK, Dickson SL: Systemic administration of ghrelin induces Fos and Egr-1 proteins in the hypothalamic arcuate nucleus of fasted and fed rats. J Neuroendocrinol 2000;12:1047–1049.
88 Horvath TL, Diano S, Sotonyi P, Heiman M, Tschop M: Minireview: ghrelin and the regulation of energy balance – a hypothalamic perspective. Endocrinology 2001;142:4163–4169.
89 Swanson LW, Sawchenko PE: Hypothalamic integration: organization of the paraventricular and supraoptic nuclei. Annu Rev Neurosci 1983;6:269–324.
90 Wellman PJ, Davies BT, Morien A, McMahon L: Modulation of feeding by hypothalamic paraventricular nucleus alpha 1- and alpha 2-adrenergic receptors. Life Sci 1993;53:669–679.
91 Shiraishi T: Noradrenergic neurons modulate lateral hypothalamic chemical and electrical stimulation-induced feeding by sated rats. Brain Res Bull 1991;27:347–351.
92 Gasparini L, Netzer WJ, Greengard P, Xu H: Does insulin dysfunction play a role in Alzheimer's disease? Trends Pharmacol Sci 2002;23:288–293.
93 Schubert M, Brazil DP, Burks DJ, Kushner JA, Ye J, Flint CL, Farhang-Fallah J, Dikkes P, Warot XM, Rio C, Corfas G, White MF: Insulin receptor substrate-2 deficiency impairs brain growth and promotes tau phosphorylation. J Neurosci 2003;23:7084–7092.
94 Devaskar SU, Giddings SJ, Rajakumar PA, Carnaghi LR, Menon RK, Zahm DS: Insulin gene expression and insulin synthesis in mammalian neuronal cells. J Biol Chem 1994;269:8445–8454.
95 Young WS III: Periventricular hypothalamic cells in the rat brain contain insulin mRNA. Neuropeptides 1986;8:93–97.
96 Aizenman Y, de Vellis J: Brain neurons develop in a serum and glial free environment: effects of transferrin, insulin, insulin-like growth factor-I and thyroid hormone on neuronal survival, growth and differentiation. Brain Res 1987;406:32–42.
97 Schechter R, Abboud M, Johnson G: Brain endogenous insulin effects on neurite growth within fetal rat neuron cell cultures. Brain Res Dev Brain Res 1999;116:159–167.
98 Schechter R, Yanovitch T, Abboud M, Johnson G III, Gaskins J: Effects of brain endogenous insulin on neurofilament and MAPK in fetal rat neuron cell cultures. Brain Res 1998;808:270–278.
99 Bouret SG, Draper SJ, Simerly RB: Trophic action of leptin on hypothalamic neurons that regulate feeding. Science 2004;304:108–110.
100 Udagawa J, Hashimoto R, Suzuki H, Hatta T, Sotomaru Y, Hioki K, Kagohashi Y, Nomura T, Minami Y, Otani H: The role of leptin in the development of the cerebral cortex in mouse embryos. Endocrinology 2006;147:647–658.
101 Udagawa J, Nimura M, Otani H: Leptin affects oligodendroglial development in the mouse embryonic cerebral cortex. Neuro Endocrinol Lett 2006;27:177–182.
102 Kagotani Y, Hashimoto T, Tsuruo Y, Kawano H, Daikoku S, Chihara K: Development of the neuronal system containing neuropeptide Y in the rat hypothalamus. Int J Dev Neurosci 1989;7:359–374.
103 Tong Y, Dumont Y, Shen SH, Quirion R: Comparative developmental profile of the neuropeptide Y Y1 receptor gene and protein in the rat brain. Brain Res Mol Brain Res 1997;48:323–332.
104 Herrera E, Lasuncion MA, Huerta L, Martin-Hidalgo A: Plasma leptin levels in rat mother and offspring during pregnancy and lactation. Biol Neonate 2000;78:315–320.
105 Matsuda J, Yokota I, Tsuruo Y, Murakami T, Ishimura K, Shima K, Kuroda Y: Development changes in long-form leptin receptor expression and localization in rat brain. Endocrinology 1999;140:5233–5238.
106 Wirth JB, Epstein AN: Ontogeny of thirst in the infant rat. Am J Physiol 1976;230:188–198.
107 Bouret SG, Draper SJ, Simerly RB: Formation of projection pathways from the arcuate nucleus of the hypothalamus to hypothalamic regions implicated in the neural control of feeding behavior in mice. J Neurosci 2004;24:2797–2805.
108 Ahima RS, Prabakaran D, Flier JS: Postnatal leptin surge and regulation of circadian rhythm of leptin by feeding. Implications for energy homeostasis and neuroendocrine function. J Clin Invest 1998;101:1020–1027.
109 Bouret SG, Simerly RB: Developmental programming of hypothalamic feeding circuits. Clin Genet 2006;70:295–301.
110 Kohno D, Nakata M, Maekawa F, Fujiwara K, Maejima Y, Kuramochi M, Shimazaki T, Okano H, Onaka T, Yada T: Leptin suppresses ghrelin-induced activation of neuropeptide Y neurons in the arcuate nucleus via phosphatidylinositol 3-kinase- and phosphodiesterase 3-mediated pathway. Endocrinology 2007;148:2251–2263.
111 Cui H, Cai F, Belsham DD: Leptin signaling in neurotensin neurons involves STAT, MAP kinases ERK1/2, and p38 through c-Fos and ATF1. FASEB J 2006;20:2654–2656.
112 Davidowa H, Plagemann A: Decreased inhibition by leptin of hypothalamic arcuate neurons in neonatally overfed young rats. Neuroreport 2000;11:2795–2798.

113 Suburo AM, Gibson SJ, Moscoso G, Terenghi G, Polak JM: Transient expression of neuropeptide Y and its C-flanking peptide immunoreactivities in the spinal cord and ganglia of human embryos and fetuses. Neurosci 1992;46:571–584.
114 Kawamura K, Takebayashi S: The development of noradrenaline-, acetylcholinesterase-, neuropeptide Y- and vasoactive intestinal polypeptide-containing nerves in human cerebral arteries. Neurosci Lett 1994;175:1–4.
115 Koutcherov Y, Mai JK, Ashwell KW, Paxinos G: Organization of human hypothalamus in fetal development. J Comp Neurol 2002;446:301–324.
116 Makino Y, Hosoda H, Shibata K, Makino I, Kojima M, Kangawa K, Kawarabayashi T: Alteration of plasma ghrelin levels associated with the blood pressure in pregnancy. Hypertension 2002;39:781–784.
117 Cetin I, Morpurgo PS, Radaelli T, Taricco E, Cortelazzi D, Bellotti M, Pardi G, Beck-Peccoz P: Fetal plasma leptin concentrations: relationship with different intrauterine growth patterns from 19 weeks to term. Pediatr Res 2000;48:646–651.
118 Jaquet D, Leger J, Levy-Marchal C, Oury JF, Czernichow P: Ontogeny of leptin in human fetuses and newborns: effect of intrauterine growth retardation on serum leptin concentrations. J Clin Endocrinol Metab 1998;83:1243–1246.
119 Matsuda J, Yokota I, Iida M, Murakami T, Yamada M, Saijo T, Naito E, Ito M, Shima K, Kuroda Y: Dynamic changes in serum leptin concentrations during the fetal and neonatal periods. Pediatr Res 1999;45:71–75.
120 Cinaz P, Sen E, Bideci A, Ezgu FS, Atalay Y, Koca E: Plasma leptin levels of large for gestational age and small for gestational age infants. Acta Paediatr 1999; 88:753–756.
121 Akcurin S, Velipasaoglu S, Akcurin G, Guntekin M: Leptin profile in neonatal gonadotropin surge and relationship between leptin and body mass index in early infancy. J Pediatr Endocrinol Metab 2005;18: 189–195.
122 Singer LK, Kuper J, Brogan RS, Smith MS, Grove KL: Novel expression of hypothalamic neuropeptide Y during postnatal development in the rat. Neuroreport 2000;11:1075–1080.
123 Varma A, He J, Weissfeld L, Devaskar SU: Postnatal intracerebroventricular exposure to neuropeptide Y causes weight loss in female adult rats. Am J Physiol Regul Integr Comp Physiol 2003;284:R1560–R1566.
124 El-Haddad MA, Ismail Y, Day L, Ross MG: Unopposed appetite (orexigenic) mechanisms in the near-term ovine fetus: central leptin does not inhibit sucrose ingestion. J Matern Fetal Neonatal Med 2004;15:291–296.
125 Roberts TJ, Nijland MJ, Caston-Balderrama A, Ross MG: Central leptin stimulates ingestive behavior and urine flow in the near term ovine fetus. Horm Metab Res 2001;33:144–150.
126 Lin J, Richard BC, Kraeling RR, Rampacek GB: Developmental changes in the long form leptin receptor and related neuropeptide gene expression in the pig brain. Biol Reprod 2001;64:1614–1618.
127 Proulx K, Richard D, Walker CD: Leptin regulates appetite-related neuropeptides in the hypothalamus of developing rats without affecting food intake. Endocrinology 2002;143:4683–4692.
128 Mistry AM, Swick A, Romsos DR: Leptin alters metabolic rates before acquisition of its anorectic effect in developing neonatal mice. Am J Physiol 1999;277:R742–R747.
129 Warnes KE, Morris MJ, Symonds ME, Phillips ID, Clarke IJ, Owens JA, McMillen IC: Effects of increasing gestation, cortisol and maternal undernutrition on hypothalamic neuropeptide Y expression in the sheep fetus. J Neuroendocrinol 1998; 10:51–57.
130 Plagemann A, Waas T, Harder T, Rittel F, Ziska T, Rohde W: Hypothalamic neuropeptide Y levels in weaning offspring of low-protein malnourished mother rats. Neuropeptides 2000;34:1–6.
131 Yousheng J, Nguyen T, Desai M, Ross MG: Programmed alterations in hypothalamic neuronal orexigenic responses to ghrelin following gestational nutrient restriction. Reprod Sci 2008;15:702–709.
132 Fahrenkrog S, Harder T, Stolaczyk E, Melchior K, Franke K, Dudenhausen JW, Plagemann A: Cross-fostering to diabetic rat dams affects early development of mediobasal hypothalamic nuclei regulating food intake, body weight, and metabolism. J Nutr 2004;134:648–654.
133 Franke K, Harder T, Aerts L, Melchior K, Fahrenkrog S, Rodekamp E, Ziska T, Van Assche FA, Dudenhausen JW, Plagemann A: 'Programming' of orexigenic and anorexigenic hypothalamic neurons in offspring of treated and untreated diabetic mother rats. Brain Res 2005;1031:276–283.
134 Kozak R, Burlet A, Burlet C, Beck B: Dietary composition during fetal and neonatal life affects neuropeptide Y functioning in adult offspring. Brain Res Dev Brain Res 2000;125:75–82.
135 Ferezou-Viala J, Roy AF, Serougne C, Gripois D, Parquet M, Bailleux V, Gertler A, Delplanque B, Djiane J, Riottot M, Taouis M: Long-term consequences of maternal high-fat feeding on hypothalamic leptin sensitivity and diet-induced obesity in the offspring. Am J Physiol Regul Integr Comp Physiol 2007;293:R1056–R1062.

136 Geary M, Pringle PJ, Persaud M, Wilshin J, Hindmarsh PC, Rodeck CH, Brook CG: Leptin concentrations in maternal serum and cord blood: relationship to maternal anthropometry and fetal growth. Br J Obstet Gynaecol 1999;106:1054–1060.
137 Hoggard N, Haggarty P, Thomas L, Lea RG: Leptin expression in placental and fetal tissues: does leptin have a functional role? Biochem Soc Trans 2001;29:57–63.
138 Blum JW, Zbinden Y, Hammon HM, Chilliard Y: Plasma leptin status in young calves: effects of pre-term birth, age, glucocorticoid status, suckling, and feeding with an automatic feeder or by bucket. Domest Anim Endocrinol 2005;28:119–133.
139 Kotani Y, Yokota I, Kitamura S, Matsuda J, Naito E, Kuroda Y: Plasma adiponectin levels in newborns are higher than those in adults and positively correlated with birth weight. Clin Endocrinol (Oxf) 2004;61:418–423.
140 McMillen IC, Muhlhausler BS, Duffield JA, Yuen BS: Prenatal programming of postnatal obesity: fetal nutrition and the regulation of leptin synthesis and secretion before birth. Proc Nutr Soc 2004;63:405–412.
141 Eckert JE, Gatford KL, Luxford BG, Campbell RG, Owens PC: Leptin expression in offspring is programmed by nutrition in pregnancy. J Endocrinol 2000;165:R1–R6.
142 Singhal A, Farooqi IS, O'Rahilly S, Cole TJ, Fewtrell M, Lucas A: Early nutrition and leptin concentrations in later life. Am J Clin Nutr 2002;75:993–999.
143 He Q, Karlberg J: Prediction of adult overweight during the pediatric years. Pediatr Res 1999;46:697–703.
144 Gasser T, Ziegler P, Seifert B, Molinari L, Largo RH, Prader A: Prediction of adult skinfolds and body mass from infancy through adolescence. Ann Hum Biol 1995;22:217–233.
145 Mott GE, Lewis DS, McGill HC: Programming of metabolism by breast or formula feeding. Ciba Found Symp 1999;156:56–76.
146 De Blasio MJ, Gatford KL, Robinson JS, Owens JA: Placental restriction of fetal growth reduces size at birth and alters postnatal growth, feeding activity, and adiposity in the young lamb. Am J Physiol Regul Integr Comp Physiol 2007;292: R875–R886.
147 Dulloo AG: Thrifty energy metabolism in catch-up growth trajectories to insulin and leptin resistance. Best Pract Res Clin Endocrinol Metab 2008;22:155–171.
148 Casillas E, Desai M, Han G, Tosh DN, Ross MG: Suppressed hypothalamic insulin signaling via PI3K and MAPK in IUGR newborns: mechanism of impaired anorexigenic responses and programmed obesity. Reprod Sci 2008;15:196A.
149 Desai M, Casillas E, Han G, Tosh DN, Ross MG: Programmed alteration of hypothalamic leptin and insulin signaling pathways contributes to reduced anorexigenic responses in IUGR offspring. Reprod Sci 2008;15:192A.
150 Sardinha FL, Telles MM, Albuquerque KT, Oyama LM, Guimaraes PA, Santos OF, Ribeiro EB: Gender difference in the effect of intrauterine malnutrition on the central anorexigenic action of insulin in adult rats. Nutrition 2006;22:1152–1161.
151 Albuquerque KT, Sardinha FL, Telles MM, Watanabe RL, Nascimento CM, Tavares do Carmo MG, Ribeiro EB: Intake of trans fatty acid-rich hydrogenated fat during pregnancy and lactation inhibits the hypophagic effect of central insulin in the adult offspring. Nutrition 2006;22:820–829.
152 Irani BG, Dunn-Meynell AA, Levin BE: Altered hypothalamic leptin, insulin, and melanocortin binding associated with moderate-fat diet and predisposition to obesity. Endocrinology 2007;148:310–316.
153 Clegg DJ, Benoit SC, Reed JA, Woods SC, Dunn-Meynell A, Levin BE: Reduced anorexic effects of insulin in obesity-prone rats fed a moderate-fat diet. Am J Physiol Regul Integr Comp Physiol 2005;288: R981–R986.
154 Passos MC, Vicente LL, Lisboa PC, de Moura EG: Absence of anorectic effect to acute peripheral leptin treatment in adult rats whose mothers were malnourished during lactation. Horm Metab Res 2004;36:625–629.
155 Martin RL, Perez E, He YJ, Dawson R Jr, Millard WJ: Leptin resistance is associated with hypothalamic leptin receptor mRNA and protein downregulation. Metab 2000;49:1479–1484.
156 Gluckman PD, Lillycrop KA, Vickers MH, Pleasants AB, Phillips ES, Beedle AS, Burdge GC, Hanson MA: Metabolic plasticity during mammalian development is directionally dependent on early nutritional status. Proc Natl Acad Sci U S A 2007;104: 12796–12800.
157 He B, You L, Uematsu K, Zang K, Xu Z, Lee AY, Costello JF, McCormick F, Jablons DM: SOCS-3 is frequently silenced by hypermethylation and suppresses cell growth in human lung cancer. Proc Natl Acad Sci U S A 2003;100:14133–14138.

158 Vickers MH, Gluckman PD, Coveny AH, Hofman PL, Cutfield WS, Gertler A, Breier BH, Harris M: Neonatal leptin treatment reverses developmental programming. Endocrinol 2005;146:4211–4216.

159 Toste FP, de Moura EG, Lisboa PC, Fagundes AT, de OE, Passos MC: Neonatal leptin treatment programmes leptin hypothalamic resistance and intermediary metabolic parameters in adult rats. Br J Nutr 2006;95:830–837.

Dr. Mina Desai
Department of Obstetrics and Gynecology
Harbor-UCLA Medical Center, Geffen School of Medicine at UCLA
1024 W. Carson St., RB-1, Box 446, Los Angeles, CA 90502 (USA)
Tel. +1 310 222 1961, Fax +1 310 222 4131, E-Mail mdesai@obgyn.humc.edu

Newnham JP, Ross MG (eds): Early Life Origins of Human Health and Disease.
Basel, Karger, 2009, pp 164–173

Fetal Programming of Type 2 Diabetes

Ana Piekarz · Susan E. Ozanne

Institute of Metabolic Science, Addenbrooke's Hospital, University of Cambridge,
Cambridge, UK

Fetal growth is a complex and dynamic process in which adequate nutrition is vital. From the beginning of pregnancy, maternal nutrition and placental function are essential to the growth of a healthy fetus. The exchange of nutrients, oxygen and endocrine signals is especially critical. Growth potential depends on the coordinated interaction of the maternal-placental-fetal units. Perturbations of these units will have an impact on fetal growth with major implications for developmental programming [1]. The term 'programming' describes a process where a stimulus or insult during a critical period of growth and development results in long-term or permanent effects on the structure or function of an organ or system [2].

There is strong epidemiological evidence that hypertension, insulin resistance and dyslipidaemia, leading to increased rates of cardiovascular disease and type 2 diabetes in adult life, originate in early life. The 'thrifty phenotype hypothesis' proposes that fetal metabolic adaptations to poor nutrition, caused by maternal malnutrition or placental insufficiency, promote postnatal survival by selecting an appropriate trajectory of growth. If the growing fetus experiences deficient nutrient supply, it will maximize metabolic efficiency and food searching behaviour in order to protect tissues important for immediate survival, such as the brain. Although these adaptations may be beneficial for short-term survival, they can lead to permanent alterations in metabolism, body structure and physiology, therefore leading to increased risk of developing chronic diseases later in life in the face of adequate or excess nutrition.

Expanding on the thrifty phenotype hypothesis, the 'predictive adaptative response hypothesis' proposes that the fetus makes adaptations not only in utero but also during early postnatal life, anticipating its future environment and preparing the organism for survival. If there is a mismatch between the actual environment and the one the fetus has predicted, the growth trajectory becomes inappropriate, leading to increased risk of disease [3].

In addition to these environmentally based hypotheses, it has also been suggested that these relationships have a genetic basis. Fetal insulin is one of the key determinants of fetal growth. The 'fetal insulin hypothesis' suggests that genetically determined defects in insulin secretion or insulin action result in impaired insulin-mediated growth of the fetus, low birthweight and increased susceptibility to diabetes in adult life [4]. Evidence supporting the fetal insulin hypothesis comes from monogenic forms of diabetes. Fetal mutations in the glucokinase gene lead to reduced birthweight and the development of type 2 diabetes at maturity [5].

Epidemiological Evidence

Extensive data from human epidemiological studies have provided strong evidence of the association between intrauterine growth restriction and predisposition to metabolic diseases. This association was initially observed in relation to cardiovascular disease, and subsequently extended to cardiovascular risk factors. The first published study linking birthweight and diabetes examined a cohort of men in Hertfordshire, UK [6]. At the age of 64 years, men with low birthweight (<2.5 kg) were more likely to have impaired glucose tolerance and type 2 diabetes compared with those with normal or high birthweight. This research group later found a similar relationship in another cohort of English men and women [7]. The trend observed was independent of current body mass index and social class. Since then, these original observations have been replicated in numerous populations [8]. A variation on these findings is the U-shaped association of birthweight and diabetes in Pima Indians, which is also related to gestational diabetes. A higher prevalence of diabetes was observed in subjects with both low and high birthweights [9].

Populations exposed in utero to periods of undernutrition during World War II have been examined. Studies of children born during the Dutch Hunger Winter have shown that poor maternal nutrition during the last trimester of pregnancy resulted in growth retardation of the fetus. Moreover, later in life an increase in plasma glucose concentrations was shown 2 h after a standard glucose test [10].

Hence, epidemiological studies provide evidence of the association between prenatal environment, birthweight and type 2 diabetes in adulthood. Further, epidemiological studies indicate that prenatal undernutrition or overnutrition may impair glucose tolerance in adulthood.

Animal Models

Epidemiological studies can only provide limited information on the underlying biological mechanisms responsible for the association between intrauterine growth restriction and the risk of developing diabetes later in life. Therefore, experimental animal models of intrauterine growth restriction and subsequent low birthweight have been developed to elucidate fundamental mechanisms. Although some studies have been conducted in large species such as sheep and pigs, rodent models have been the most extensively used. Rodents present shorter gestation and lifespan compared with larger species. This results in less costly, shorter and highly informative studies. Rodent models of growth restriction have been shown to successfully programme the endocrine pancreas and insulin-sensitive tissues such as pancreas, muscle and kidneys.

Dietary Manipulation

Low Protein

The low protein model is one of the most extensively studied. This model generally consists of feeding rodent dams a low-protein diet (8%) during pregnancy and lactation in order to induce growth restriction in the offspring. These are compared with offspring born to mothers fed a control protein diet (20%). Offspring of low protein dams show reduced birthweight and a wide range of metabolic disturbances. Studies conducted by our group have shown that the maternal low protein offspring have improved glucose tolerance at an early age (12 weeks), followed by a subsequent age-dependent decline in glucose tolerance at 15 months and type 2 diabetes and insulin resistance by 17 months [11–13].

Specific changes in insulin signalling protein expression were observed in the low protein rat offspring [14]. Striking parallels in the expression profile of insulin signalling components, between the low protein rat model and human tissue were observed, both in terms of specificity and magnitude of the effect. GLUT 4, protein kinase C zeta and the p85 regulatory subunit of phosphatidylinositol-3-kinase were reduced in muscle from both low birthweight humans [15] and the low protein rat model [16]. There were no changes in the expression of glycogen synthetase kinase and Akt.

The pancreatic beta cell mass is determined by number and size. This unique cell population is required for insulin production to maintain glucose homeostasis. Several experimental studies have investigated the long-term effects of maternal protein restriction on early beta cell development. Reduced insulin secretion in low protein pups may result from one or more intrinsic alterations, such as smaller size, reduced number of beta cells [17], reduced proliferation of islet cells, reduced islet vascularization [18, 19] or increased beta cell apoptosis [20].

Total Calorie Restriction

A number of studies have addressed the timing and effects of total energy restriction during prenatal and early postnatal life in the development of the endocrine pancreas. Early studies showed that acute protein-calorie restriction between 2–6 weeks of age impairs glucose tolerance and insulin secretion in rats [21, 22]. More recent studies have focused on the long-term effect of maternal calorie restriction. Maternal food restriction to 50% ad lib during late pregnancy decreases beta cell mass in the offspring at birth [23]. In contrast to the low protein effect, the reduced beta cell mass was not attributed to lower proliferation or increased apoptosis but rather to alteration in islet neogenesis, suggesting that the mechanism involved in the reduced beta cell mass may be different.

The reduction in beta cell mass and number was not restored in adult age despite normal nutrition after weaning. [23]. Age-dependent glucose intolerance was observed in animals with decreased beta cell mass. Beta cell mass does not expand and apoptosis increases with age [24]. Fetal and early postnatal malnutrition also impairs the adaptation of the endocrine pancreas to the increased insulin demand of a subsequent pregnancy [25]. Beta cell dysfunction was also observed in the next generation [26].

A more severe food restriction (30% of ad lib intake) throughout pregnancy has been shown to result in systolic hypertension, hyperinsulinaemia, hyperleptinaemia and obesity in the offspring [27]. Treatment with IGF-1 as an adult resulted in normalization of blood pressure and also reduction in plasma insulin and leptin concentrations [28]. In guinea pigs, maternal food restriction reduces fetal growth and causes hyperinsulinaemia in adult offspring [29].

Iron Restriction

The importance of adequate iron intake during pregnancy is well established. The effects of dietary-induced iron deficiency on fetal and maternal metabolism have been studied in rats. Iron deficiency during pregnancy leads to problems for the mother and the developing fetus [30]. Maternal iron restriction in the rat has been reported to decrease birthweight, and maternal anaemia has been shown to result in increased blood pressure and decreased serum lipids in 3-month-old offspring [31, 32]. Systolic blood pressure was also elevated in the offspring of iron-restricted dams at 16 months of age. However, no differences were observed in glucose tolerance between iron restricted and control groups at this age [32].

High-Fat Diet

In the Pima Indians, a population with very high rates of gestational diabetes, the association between birthweight and diabetes is U-shaped. The highest prevalence of diabetes occurs in both high and low birthweight individuals [9]. These findings suggest that prenatal overnutrition can also programme adult disease.

Offspring of rats fed high saturated fat diets during pregnancy have fetal insulin resistance [33]. Furthermore, exposure to a diet rich in animal fat throughout pregnancy and lactation results in endothelial dysfunction [34], hyperlipidaemia and increased adiposity [35]. At 3 months of age, male offspring from dams fed a diet rich in omega-6 polyunsaturated fats showed normal insulin sensitivity and glucose tolerance, although they were hyperinsulinaemic after an oral glucose challenge. Insulin signalling protein expression levels were consistent with reduced hepatic insulin sensitivity [36].

Pharmacological Manipulation

Glucocorticoid Exposure

Prenatal glucocorticoid exposure or stress has been suggested as a potential mechanism linking fetal growth with adult disease. Prenatal treatment with glucocorticoids slows fetal growth and results in low birthweight [37]. Systolic blood pressure [38] and blood glucose levels [39] rise in the adult offspring after prenatal exposure to glucocorticoids. In rats, exposure to glucocorticoids during the last week of pregnancy results in permanent upregulation of hepatic metabolic enzyme expression and activity, which contributes to the glucose intolerance seen in the offspring. The increase in phosphoenolpyruvate carboxykinase 2 (PCK2), the rate-controlling enzyme of gluconeogenesis, is accompanied by an increase in hepatocyte nuclear factor 4α (HNF4α) mRNA. These data suggest that HNF4α might mediate PCK2 overexpression and subsequent hyperglycaemia [40]. Other effects observed include increased hypothalamic-pituitary-adrenal axis activity and tissue sensitivity to glucocorticoid hormones. It has been suggested that these effects may be due to physiological variations in placental 11β hydroxysteroid dehydrogenase type 2, the enzyme responsible for catalysing the metabolism of active physiological glucocorticoids to inert 11-keto forms [41].

Maternal Diabetes

Gestational diabetes affects endocrine pancreas development in both humans and animals [42]. In rats, gestational diabetes induced by administration of streptozotocin (STZ) has been used to elucidate the effects of diabetes on fetal growth and development. STZ is a chemical that selectively destroys beta cells of the pancreatic islets, resulting in mild or severe diabetes depending on the dose used. A low dose of STZ results in mild gestational diabetes resulting in increased blood glucose concentration, hyperinsulinaemia and fetal macrosomia, whereas a high dose induces insulin deficiency diabetes-associated growth restriction [43, 44]. Offspring of both the mild and severe diabetic STZ models subsequently develop gestational diabetes.

Surgical Models

Surgical interventions have been used to generate further evidence of phenomena found in dietary or pharmacological models of programming.

Uterine Artery Ligation
Unilateral and bilateral artery ligation has been used to generate placental insufficiency, a common cause of intrauterine growth restriction [45]. Growth-retarded offspring have a reduced number of nephrons, which impairs renal function [46] despite a compensatory glomerular enlargement associated with an increased proteinuria [47]. This model of growth restriction has been associated with reduced beta cell mass [48], mild insulin resistance and beta cell secretory defects early in life [49]. The level of Pdx1 mRNA, a regulator of beta cell proliferation, was significantly decreased in fetal beta cells in this model. Early programming of the endocrine pancreas after uterine artery ligation led to altered glucose tolerance later in life and the development of type 2 diabetes at 26 weeks of age [50]. Diabetes has been shown to be transmitted to the second generation [51].

Transgenerational Effect

Although most attention has been given to the first generation of offspring exposed to an altered intrauterine environment, there is evidence that consequences extend to future generations. In addition to the effects of the mothers' dietary intake on birthweight, mothers' birthweights are related to those of their children and grandchildren [52]. During pregnancy, the fetus is completely dependent on nutrients supplied by the mother. Thus, fetal development in an altered intrauterine environment can induce alterations in fetal metabolism, with lasting consequences for the glucose tolerance of the offspring in adult life. The second generation also features a diabetogenic tendency [53, 54]. Using embryo transfer methodology, it was shown that the presence of altered glucose/insulin homeostasis and muscle insulin signalling is independent of the intermediate intrauterine metabolic environment [55]. In rats, the effects of prenatal glucocorticoid exposure are not only observed in the immediate offspring as adults, but are also present in the second-generation offspring [56].

Mechanisms in Humans

Several mechanistic studies in humans have focused on insulin action. These studies have benefited from the substantial progress that has been made in understanding how insulin mediates its effects. Through a combination of approaches, the way insulin signals its metabolic actions has been determined in detail. The insulin receptor is well studied and the mechanisms that mediate insulin action on insulin-responsive glucose transport are well characterized [57].

Insulin signalling molecules have been studied in tissue biopsies from low birthweight subjects. Those men with low birthweight (bottom 10th percentile) had increased fasting plasma glucose and increased 2-hour glucose levels during an oral glucose challenge [58]. Furthermore, a case-control cohort of 1,621 subjects showed that those born small for gestational age are more likely to develop insulin resistance in adulthood [59].

Molecular Mechanisms

Despite the significant number of studies published, the mechanisms responsible for intrauterine programming of type 2 diabetes are not completely understood. Mechanistic pathways that have been studied include DNA methylation of key genes [60, 61], altered gene expression in specific tissues and programming of the hypothalamic-pituitary-adrenal axis. As with many complex diseases, it cannot be assumed that a unifying pathway links the intrauterine environment and chronic disease later in life, suggesting that the interactions between these mechanistic pathways also need to be established.

Future Directions

Scientists working in this field are faced with three primary challenges. Firstly, more detailed animal studies are needed to uncover the specific pathways and molecular mechanisms by which altered intrauterine environment leads to glucose intolerance and type 2 diabetes. Secondly, future studies need to further explore the period of susceptibility, including the pre-conception period, during which nutritional perturbations produce long-term consequences and the fundamental pathways by which it arises. Thirdly, birthweight, a crude index of intrauterine growth restriction, is not sufficient to predict if a person will develop adult-onset disease. Thus, another key area for future research is identifying early markers indicative of suboptimal fetal experience. The significance of obtaining this knowledge cannot be overstated. Only when such information is available, will it be possible to devise methods for primary intervention.

References

1 Fowden AL, Ward JW, Wooding FP, Forhead AJ, Constancia M: Programming placental nutrient transport capacity. J Physiol 2006;572:5–15.

2 Lucas A: Programming by early nutrition in man. Ciba Found Symp 1991;156:38–50; discussion 50–35.

3 Gluckman PD, Hanson MA, Spencer HG: Predictive adaptive responses and human evolution. Trends Ecol Evol 2005;20:527–533.

4 Hattersley AT, Tooke JE: The fetal insulin hypothesis: an alternative explanation of the association of low birthweight with diabetes and vascular disease. Lancet 1999;353:1789–1792.

5 Stride A, Hattersley AT: Different genes, different diabetes: lessons from maturity-onset diabetes of the young. Ann Med 2002;34:207–216.

6 Hales CN, Barker DJ, Clark PM, Cox LJ, Fall C, Osmond C, Winter PD: Fetal and infant growth and impaired glucose tolerance at age 64. BMJ 1991;303: 1019–1022.
7 Phipps K, Barker DJ, Hales CN, Fall CH, Osmond C, Clark PM: Fetal growth and impaired glucose tolerance in men and women. Diabetologia 1993;36: 225–228.
8 Hales CN, Barker DJ: The thrifty phenotype hypothesis. Br Med Bull 2001;60:5–20.
9 McCance DR, Pettitt DJ, Hanson RL, Jacobsson LT, Knowler WC, Bennett PH: Birth weight and non-insulin dependent diabetes: thrifty genotype, thrifty phenotype, or surviving small baby genotype? BMJ 1994;308:942–945.
10 Ravelli AC, van der Meulen JH, Michels RP, Osmond C, Barker DJ, Hales CN, Bleker OP: Glucose tolerance in adults after prenatal exposure to famine. Lancet 1998;351:173–177.
11 Ozanne SE, Nave BT, Wang CL, Shepherd PR, Prins J, Smith GD: Poor fetal nutrition causes long-term changes in expression of insulin signaling components in adipocytes. Am J Physiol 1997;273:E46–E51.
12 Ozanne SE, Wang CL, Coleman N, Smith GD: Altered muscle insulin sensitivity in the male offspring of protein malnourished rats. Am J Physiol 1996;271:E1128–E1134.
13 Petry CJ, Dorling MW, Pawlak DB, Ozanne SE, Hales CN: Diabetes in old male offspring of rat dams fed a reduced protein diet. Int J Exp Diabetes Res 2001;2:139–143.
14 Ozanne SE, Olsen GS, Hansen LL, Tingey KJ, Nave BT, Wang CL, Hartil K, Petry CJ, Buckley AJ, Mosthaf-Seedorf L: Early growth restriction leads to down regulation of protein kinase c zeta and insulin resistance in skeletal muscle. J Endocrinol 2003;177: 235–241.
15 Ozanne SE, Jensen CB, Tingey KJ, Storgaard H, Madsbad S, Vaag AA: Low birthweight is associated with specific changes in muscle insulin-signalling protein expression. Diabetologia 2005;48:547–552.
16 Ozanne SE, Dorling MW, Wang CL, Nave BT: Impaired pi 3-kinase activation in adipocytes from early growth-restricted male rats. Am J Physiol Endocrinol Metab 2001;280:E534–E539.
17 Dahri S, Snoeck A, Reusens-Billen B, Remacle C, Hoet JJ: Islet function in offspring of mothers on low-protein diet during gestation. Diabetes 1991; (suppl 2):115–120.
18 Snoeck A, Remacle C, Reusens B, Hoet JJ: Effect of a low protein diet during pregnancy on the fetal rat endocrine pancreas. Biol Neonate 1990;57:107–118.
19 Boujendar S, Reusens B, Merezak S, Ahn MT, Arany E, Hill D, Remacle C: Taurine supplementation to a low protein diet during foetal and early postnatal life restores a normal proliferation and apoptosis of rat pancreatic islets. Diabetologia 2002;45:856–866.
20 Petrik J, Reusens B, Arany E, Remacle C, Coelho C, Hoet JJ, Hill DJ: A low protein diet alters the balance of islet cell replication and apoptosis in the fetal and neonatal rat and is associated with a reduced pancreatic expression of insulin-like growth factor-II. Endocrinology 1999;140:4861–4873.
21 Swenne I, Crace CJ, Milner RD: Persistent impairment of insulin secretory response to glucose in adult rats after limited period of protein-calorie malnutrition early in life. Diabetes 1987;36:454–458.
22 Swenne I, Crace CJ, Jansson L: Intermittent protein-calorie malnutrition in the young rat causes long-term impairment of the insulin secretory response to glucose in vitro. J Endocrinol 1988;118:295–302.
23 Garofano A, Czernichow P, Breant B: Beta-cell mass and proliferation following late fetal and early postnatal malnutrition in the rat. Diabetologia 1998;41: 1114–1120.
24 Garofano A, Czernichow P, Breant B: Effect of ageing on beta-cell mass and function in rats malnourished during the perinatal period. Diabetologia 1999;42:711–718.
25 Blondeau B, Garofano A, Czernichow P, Breant B: Age-dependent inability of the endocrine pancreas to adapt to pregnancy: a long-term consequence of perinatal malnutrition in the rat. Endocrinology 1999;140:4208–4213.
26 Blondeau B, Avril I, Duchene B, Breant B: Endocrine pancreas development is altered in foetuses from rats previously showing intra-uterine growth retardation in response to malnutrition. Diabetologia 2002;45:394–401.
27 Vickers MH, Breier BH, Cutfield WS, Hofman PL, Gluckman PD: Fetal origins of hyperphagia, obesity, and hypertension and postnatal amplification by hypercaloric nutrition. Am J Physiol Endocrinol Metab 2000;279:E83–E87.
28 Vickers MH, Ikenasio BA, Breier BH: IGF-I treatment reduces hyperphagia, obesity, and hypertension in metabolic disorders induced by fetal programming. Endocrinology 2001;142:3964–3973.
29 Kind KL, Clifton PM, Grant PA, Owens PC, Sohlstrom A, Roberts CT, Robinson JS, Owens JA: Effect of maternal feed restriction during pregnancy on glucose tolerance in the adult guinea pig. Am J Physiol Regul Integr Comp Physiol 2003;284: R140–R152.

30 Gambling L, Danzeisen R, Fosset C, Andersen HS, Dunford S, Srai SK, HJ MC: Iron and copper interactions in development and the effect on pregnancy outcome. J Nutr 2003;133:1554S–1556S.
31 Lewis RM, Petry CJ, Ozanne SE, Hales CN: Effects of maternal iron restriction in the rat on blood pressure, glucose tolerance, and serum lipids in the 3-month-old offspring. Metabolism 2001;50:562–567.
32 Lewis RM, Forhead AJ, Petry CJ, Ozanne SE, Hales CN: Long-term programming of blood pressure by maternal dietary iron restriction in the rat. Br J Nutr 2002;88:283–290.
33 Guo F, Jen KL: High-fat feeding during pregnancy and lactation affects offspring metabolism in rats. Physiol Behav 1995;57:681–686.
34 Khan F, Elherik K, Bolton-Smith C, Barr R, Hill A, Murrie I, Belch JJ: The effects of dietary fatty acid supplementation on endothelial function and vascular tone in healthy subjects. Cardiovasc Res 2003; 59:955–962.
35 Khan IY, Dekou V, Douglas G, Jensen R, Hanson MA, Poston L, Taylor PD: A high-fat diet during rat pregnancy or suckling induces cardiovascular dysfunction in adult offspring. Am J Physiol Regul Integr Comp Physiol 2005;288:R127–R133.
36 Buckley AJ, Keseru B, Briody J, Thompson M, Ozanne SE, Thompson CH: Altered body composition and metabolism in the male offspring of high fat-fed rats. Metabolism 2005;54:500–507.
37 Reinisch JM, Simon NG, Karow WG, Gandelman R: Prenatal exposure to prednisone in humans and animals retards intrauterine growth. Science 1978; 202:436–438.
38 Benediktsson R, Lindsay RS, Noble J, Seckl JR, Edwards CR: Glucocorticoid exposure in utero: new model for adult hypertension. Lancet 1993;341:339–341.
39 Lindsay RS, Lindsay RM, Waddell BJ, Seckl JR: Prenatal glucocorticoid exposure leads to offspring hyperglycaemia in the rat: studies with the 11 beta-hydroxysteroid dehydrogenase inhibitor carbenoxolone. Diabetologia 1996;39:1299–1305.
40 Nyirenda MJ, Dean S, Lyons V, Chapman KE, Seckl JR: Prenatal programming of hepatocyte nuclear factor 4alpha in the rat: a key mechanism in the 'foetal origins of hyperglycaemia'? Diabetologia 2006;49:1412–1420.
41 Seckl JR, Meaney MJ: Glucocorticoid programming. Ann N Y Acad Sci 2004;1032:63–84.
42 Holemans K, Aerts L, Van Assche FA: Fetal growth restriction and consequences for the offspring in animal models. J Soc Gynecol Investig 2003;10:392–399.
43 Holemans K, Aerts L, Van Assche FA: Absence of pregnancy-induced alterations in tissue insulin sensitivity in the offspring of diabetic rats. J Endocrinol 1991;131:387–393.
44 Lopez-Soldado I, Herrera E: Different diabetogenic response to moderate doses of streptozotocin in pregnant rats, and its long-term consequences in the offspring. Exp Diabesity Res 2003;4:107–118.
45 Wigglesworth JS: Fetal growth retardation. Animal model: uterine vessel ligation in the pregnant rat. Am J Pathol 1974;77:347–350.
46 Merlet-Benichou C, Gilbert T, Muffat-Joly M, Lelievre-Pegorier M, Leroy B: Intrauterine growth retardation leads to a permanent nephron deficit in the rat. Pediatr Nephrol 1994;8:175–180.
47 Schreuder MF, Nyengaard JR, Fodor M, van Wijk JA, Delemarre-van de Waal HA: Glomerular number and function are influenced by spontaneous and induced low birth weight in rats. J Am Soc Nephrol 2005;16:2913–2919.
48 De Prins FA, Van Assche FA: Intrauterine growth retardation and development of endocrine pancreas in the experimental rat. Biol Neonate 1982;41:16–21.
49 Vuguin P, Raab E, Liu B, Barzilai N, Simmons R: Hepatic insulin resistance precedes the development of diabetes in a model of intrauterine growth retardation. Diabetes 2004;53:2617–2622.
50 Simmons RA, Templeton LJ, Gertz SJ: Intrauterine growth retardation leads to the development of type 2 diabetes in the rat. Diabetes 2001;50:2279–2286.
51 Boloker J, Gertz SJ, Simmons RA: Gestational diabetes leads to the development of diabetes in adulthood in the rat. Diabetes 2002;51:1499–1506.
52 Klebanoff MA, Graubard BI, Kessel SS, Berendes HW: Low birth weight across generations. JAMA 1984;252:2423–2427.
53 Aerts L, Van Assche FA: Animal evidence for the transgenerational development of diabetes mellitus. Int J Biochem Cell Biol 2006;38:894–903.
54 Reusens B, Remacle C: Intergenerational effect of an adverse intrauterine environment on perturbation of glucose metabolism. Twin Res 2001;4:406–411.
55 Thamotharan M, Garg M, Oak S, Rogers LM, Pan G, Sangiorgi F, Lee PW, Devaskar SU: Transgenerational inheritance of the insulin-resistant phenotype in embryo-transferred intrauterine growth-restricted adult female rat offspring. Am J Physiol Endocrinol Metab 2007;292:E1270–E1279.
56 Drake AJ, Walker BR, Seckl JR: Intergenerational consequences of fetal programming by in utero exposure to glucocorticoids in rats. Am J Physiol Regul Integr Comp Physiol 2005;288:R34–R38.

57 Virkamaki A, Ueki K, Kahn CR: Protein-protein interaction in insulin signaling and the molecular mechanisms of insulin resistance. J Clin Invest 1999; 103:931–943.
58 Jensen CB, Storgaard H, Dela F, Holst JJ, Madsbad S, Vaag AA: Early differential defects of insulin secretion and action in 19-year-old Caucasian men who had low birth weight. Diabetes 2002;51:1271–1280.
59 Jaquet D, Chevenne D, Czernichow P, Levy-Marchal C: No evidence for a major beta-cell dysfunction in young adults born with intra-uterine growth retardation. Pediatr Diabetes 2000;1:181–185.
60 Lillycrop KA, Phillips ES, Jackson AA, Hanson MA, Burdge GC: Dietary protein restriction of pregnant rats induces and folic acid supplementation prevents epigenetic modification of hepatic gene expression in the offspring. J Nutr 2005;135:1382–1386.
61 Bogdarina I, Welham S, King PJ, Burns SP, Clark AJ: Epigenetic modification of the renin-angiotensin system in the fetal programming of hypertension. Circ Res 2007;100:520–526.

Dr. Susan E. Ozanne
Metabolic Research Laboratories
Addenbrooke's Hospital, University of Cambridge
Level 4, Institute of Metabolic Science, Box 289, Cambridge, CB2 0QQ (UK)
Tel. +44 1223 336784, Fax +44 1223 330598, E-Mail seo10@cam.ac.uk

Newnham JP, Ross MG (eds): Early Life Origins of Human Health and Disease.
Basel, Karger, 2009, pp 174–183

Developmental Origins of Diabetes: Interventional Strategies

Rebecca A. Simmons

Department of Pediatrics, Children's Hospital Philadelphia and University of Pennsylvania, Philadelphia, Pa., USA

The Beta Cell in Intrauterine Growth-Retarded Fetuses

Uteroplacental insufficiency limits availability of substrates, growth factors and hormones to the fetus and retards growth during gestation. This abnormal intrauterine milieu modifies gene expression in pluripotential and terminally differentiated cells resulting in permanent structural and functional changes in key organs such as the pancreas, liver, and muscle [1–5]. We have developed a rat model of uteroplacental insufficiency (hereafter designated as IUGR for intrauterine growth retardation) induced by bilateral uterine artery ligation at 19 days of gestation (term is 22 days). Diabetes develops in adult animals at approximately 15–26 weeks of age and is accompanied by beta cell secretory defects and insulin resistance, the salient features of most forms of type 2 diabetes in the human [5]. Beta cell mass is normal during the first few weeks of life in IUGR rats; however, by 7 weeks of age, beta cell mass is reduced compared with controls. Most importantly, the progressive decline in beta cell mass occurs weeks prior to the onset of hyperglycemia. While insulin resistance is a critical component of human type 2 diabetes, it is the failure of beta cell function and growth that determines progression to the diabetic phenotype [6]. Thus, efforts to prevent the reduction in beta cell mass associated with diabetes could potentially prevent the development of the disease.

Neonatal Exendin-4 Prevents the Development of Diabetes Mellitus in the IUGR Rat

The incretin hormone glucagon-like peptide-1 (GLP-1) promotes the expansion of pancreatic beta cell mass by stimulating neogenesis as well as proliferation of existing beta cells [7]. During the newborn period, there is a high rate of replication, neogenesis, and apoptosis resulting in extensive remodeling of the endocrine pancreas

[8]. This appears to be followed by a second wave of neogenesis around the time of weaning. After weaning, levels of neogenesis and replication fall to very low levels but do continue throughout life [9]. Therefore, the fetal and neonatal period represents a critical window of opportunity for therapies designed to enhance beta cell mass.

GLP-1 is a well-established insulinotropic agent and stimulates insulin biosynthesis and glucose-dependent insulin secretion, via increases in intracellular cAMP and calcium [7]. We have found that administration of the long-acting GLP-1 analog, exendin-4, during the early newborn period (postnatal days 1–6) completely prevents the development of diabetes in the IUGR rat [10]. The effect of exendin-4 on glucose homeostasis is permanent, with a resultant increase in the life span of IUGR animals. The early normalization of glucose tolerance is observed on postnatal day 14, when IUGR beta cell mass is still normal, suggesting that exendin-4 exerts an effect on beta cell function of the IUGR rat that is independent of its effects on beta cell mass. It is likely that exendin-4 also improves glucose tolerance in the IUGR via extra-pancreatic effects. In other animal models, exendin-4 improves insulin sensitivity, decreases food intake, increases satiety, delays gastric emptying and suppresses glucagon release [11–15]. In preliminary unpublished studies, we have found that exendin-4 given in the newborn period to IUGR rats normalizes insulin resistance at the level of the liver, and prevents the development of obesity.

Exendin-4 stimulation of insulin secretion may be mediated through stimulation of Pdx1 levels. Pdx1 is a pancreatic homeoprotein that is critical for the early development of both the endocrine and exocrine pancreas, and it mediates glucose-responsive stimulation of insulin gene transcription [16]. Recently, it has been demonstrated that a modest reduction in Pdx1 impairs mitochondrial function and generation of NADH, resulting in blunted glucose-stimulated insulin secretion [17]. This is particularly relevant as mitochondrial function is markedly abnormal in islets of IUGR animals [18]. Messenger RNA levels of Pdx1 are significantly reduced in the IUGR fetus, and expression progressively declines over time [10]. Similar to our previous studies of exendin-4 treatment of normal and diabetic mice, exendin-4 increases Pdx1 expression in IUGR pancreas such that levels are similar to those of controls. Remarkably, only 6 days of treatment with this long-acting GLP-1 analog leads to a permanent recovery of Pdx1 expression in IUGR animals. Between days 18 and 22 of gestation in the rat, beta cell mass increases nearly 14-fold [19]. This rapid expansion of beta cells does not appear to be impaired in the IUGR fetus as beta cell mass is normal at this age. Thus, despite a 50% reduction in Pdx1 levels in IUGR pancreas, beta cell mass is not affected. This is consistent with the observation that milder reductions in Pdx1 protein levels, as occurs in the *Pdx1+/–* mice, allow for the development of a normal mass of beta cells [17, 20].

In addition to a life-long normalization of glucose tolerance, we have observed a complete rescue of the progressive decline in beta cell mass that is normally observed in IUGR rats [10]. The brief period of neonatal exendin-4 normalizes beta cell replication rate in IUGR animals measured at postnatal day 14. In the normal rat, replication of existing beta cells and formation of new beta cells are substantially greater during this period than at any other time in postnatal life [8, 19]. Therefore, even a modest reduction in neonatal beta cell proliferation rates will result in a long-term reduction in beta cell mass. It is likely that increased neogenesis from islet progenitor cells also contributes to the maintenance of beta cell mass in exendin-4-treated IUGR rats [Stoffers and Simmons, unpubl. data].

Chromatin Remodeling in the IUGR Rat

A number of investigators have shown that IUGR induces permanent changes in gene expression in the offspring, implicating an epigenetic mechanism. Epigenetic modifications of the genome provide a mechanism that allows the stable propagation of gene activity states from one generation of cells to the next. Excellent reviews on this topic appear frequently, reflecting the rapid advances of knowledge in the field [21–24]. Epigenetic states can be modified by environmental factors, which may contribute to the development of abnormal phenotypes. There are at least two distinct classes of epigenetic information that can be inherited with chromosomes. One class of epigenetic control of gene expression involves changes in chromatin proteins, usually involving modifications of histone tails. In eukaryotes, DNA is assembled with histones to form the nucleosome, in which DNA is wrapped approximately two turns around an octameric complex composed of two molecules of each of the four histones H2A, H2B, H3, and H4. The amino termini of histones can be modified by acetylation, methylation, sumoylation, phosphorylation, glycosylation, and ADP ribosylation. The most common modifications involve acetylation and methylation of lysine residues in the amino termini of H3 and H4. Increased acetylation induces transcription activation, whereas decreased acetylation usually induces transcription repression. Methylation of histones is associated with both transcription repression and activation. Moreover, lysine residues can be mono-, di-, or trimethylated in vivo, thus providing an additional mechanism of regulation. Trimethylation of lysine residues is only found at active genes, whereas dimethylation occurs in both active and inactive chromatin. Several chromatin modification states are mutually reinforcing. For example, methylation of lysine 9 on histone H3 can promote DNA methylation, and CpG methylation (see below) stimulates methylation of lysine 9 on histone H3 [25]. Thus, chromatin modifications induced by adverse stimuli are self-reinforcing and can propagate.

The second class of epigenetic regulation is DNA methylation, in which a cytosine base is modified by a DNA methyltransferase at the C5 position of cytosine, a reaction that is carried out by various members of a single family of enzymes. Approximately 70% of CpG dinucleotides in human DNA are constitutively methylated, whereas most of the unmethylated CpGs are located in CpG islands. CpG islands are CG-rich sequences located near coding sequences, and serve as promoters for the associated genes. Approximately half of mammalian genes have CpG islands. Methylation of CpG sites is also maintained by DNA methyltransferases. DNA methylation is commonly associated with gene silencing and contributes to X-chromosomal inactivation, genomic imprinting, as well as transcriptional regulation of tissue-specific genes during cellular differentiation [26–28]. The methylation status of CpG islands within promoter sequences works as an essential regulatory element by modifying the binding affinity of transcription factors to DNA binding sites.

Most CpG islands remain unmethylated in normal cells; however, under some circumstances such as cancer [29–34] and oxidative stress (see below), they can become methylated de novo. This aberrant methylation is accompanied by local changes in histone modification and chromatin structure, such that the CpG island and its embedded promoter take on a repressed conformation that is incompatible with gene transcription. It is not known why particular CpG islands are susceptible to aberrant methylation. A recent study by Feltus et al. [35] suggests that there is a 'sequence signature associated with aberrant methylation'. Of major significance to type 2 diabetes is their finding that *Pdx1*, a pancreatic homeobox transcription factor, was one of only 15 CpG genes (a total of 1,749 genes with CpG islands were examined) that were methylation susceptible under conditions of increased methylation induced by overexpression of a DNA methyl transferase.

A number of studies have suggested that uteroplacental insufficiency induces epigenetic modifications in the offspring [36–39]. Genome-wide DNA hypomethylation has been found in postnatal IUGR liver, which is associated with quantity of acetylated histone H3 [36]. Hyperacetylation of histone H3 was found to be site specific and acetylation of H3 lysine-9 (H3/K9), lysine-14 (H3/K14), and lysine-18 (H3/K18) was increased at the promoters of PGC-1 and CPTI, respectively, in IUGR liver [37]. At day 21 of life, the neonatal pattern of H3 hyperacetylation persisted only in the IUGR males. Whether hyperacetylation at these sites actually causes increased transcription of PGC-1 or CPT1 and how these findings relate to a phenotype in the offspring remains to be determined.

IUGR Is Associated with Progressive Epigenetic Modifications That Silence *Pdx1*

Pdx1 plays a critical role in the early development of both endocrine and exocrine pancreas, and then in the later differentiation and function of the beta cell. As early as 24 h after the onset of growth retardation, Pdx1 mRNA levels are reduced by more than 50% in IUGR fetal rats [10]. Suppression of Pdx1 expression persists after birth and progressively declines in the IUGR animal, implicating an epigenetic mechanism. The altered metabolic milieu of the IUGR pregnancy decreases *Pdx1* transcription in the offspring by mediating a cascade of epigenetic modifications culminating in silencing of *Pdx1* [39]. The open chromatin domain marked by histone H3 and H4 acetylation at the proximal promoter of *Pdx1* is essential for transcription. Robust

Pdx1 expression in islets from control animals is coincident with the presence of acetylated histones H3 and H4 as well as trimethylated H3K4 [39]. Loss of these marks results in Pdx1 silencing, and reversal of IUGR-induced epigenetic modifications normalizes Pdx1 expression [39].

The first epigenetic mark that is modified in beta cells of IUGR animals is histone acetylation (fig. 1). Islets isolated from IUGR fetuses show a significant decrease in H3 and H4 acetylation at the proximal promoter of *Pdx1* [39]. These changes in H3 and H4 acetylation are associated with a loss of binding of USF-1 to the proximal promoter of *Pdx1*. USF-1 is a critical activator of *Pdx1* transcription and decreased binding markedly decreases *Pdx1* transcription [40, 41]. After birth, histone deacetylation progresses and is followed by a marked decrease in H3K4 trimethylation and a significant increase in dimethylation of H3K9 in IUGR islets (fig. 1). Progression of these histone modifications parallels the progressive decrease in Pdx1 expression as glucose homeostasis deteriorates and oxidative stress increases in IUGR animals. Nevertheless, in the IUGR pup (at 2 weeks of age) these silencing histone modifications alone suppress Pdx1 expression since there is no appreciable methylation in the CpG island and reversal of histone deacetylation in IUGR islets (in the presence of active beta cell replication) is sufficient to nearly normalize Pdx1 mRNA levels [39].

The initial mechanism by which IUGR silences *Pdx1* is by recruitment of corepressors, including HDAC1 and mSin3A, which catalyze histone deacetylation – the first repressive mark observed at *Pdx1* in IUGR islets [39]. Binding of these deacetylases in turn facilitates loss of trimethylation of H3K4, further repressing Pdx1 expression (fig. 1). Our observation that inhibition of HDAC activity by TSA treatment normalizes H3K4me3 levels at *Pdx1* in IUGR islets suggests that the association of HDAC1 at *Pdx1* in IUGR islets likely serves as a platform for the recruitment of a demethylase, which catalyzes demethylation of H3K4. Lysine demethylases in the Jumonji class remove H3Kme3 and H3Kme2 [reviewed in 42], while LSD1 removes H3Kme1/2 [43]. Klose et al. [44] have recently demonstrated that the retinoblastoma-binding protein RBP2 contains a JmjC domain, which can specifically demethylate H3K4me3. However, enzymes in the Jumonji class may not catalyze H3K4me3 demethylation in IUGR islets as activity of these proteins is dependent upon the presence of an iron-binding domain [45, 46] which would be inactivated under conditions of oxidative stress that are present in IUGR islets [18]. These results imply that another class of histone demethylases may exist in beta cells.

Loss of H3K4me3 is concomitant with a marked increase in H3K9me2 at *Pdx1* in 2-week-old IUGR animals, suggesting that K4 methylation precludes methylation at lysine 9. These in vivo findings support several in vitro studies showing that active chromatin states are maintained by H3K4 methylation, which opposes the lysine methylations that characterize inactive chromatin [47, 48]. Since restoration of histone acetylation by TSA treatment of IUGR islets reverses H3K9me2, this also demonstrates that histone acetylation prevents methylation of H3K9. Thus, IUGR-induced histone modifications are mutually reinforcing and interdependent [39].

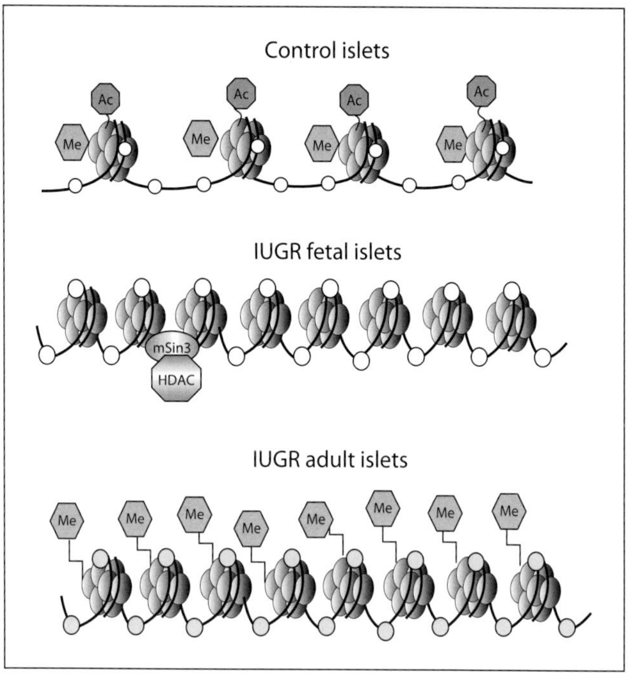

Fig. 1. Summary of epigenetic changes at *Pdx1* in IUGR rats during the development of type 2 diabetes. In pancreatic beta cells (top), the *Pdx1* proximal promoter is normally found in an unmethylated (white circles) open chromatin state allowing access to transcription factors such as USF-1 and associated with nucleosomes characterized by acetylated (Ac) histones H3 and H4 and with trimethylated H3K4 (Me). In IUGR fetal and 2-week islets (middle), histone acetylation is progressively lost through association with a mSin3A-HDAC1-DNMT1 repressor complex, with trimethylated H3K4 disappearing and dimethylated H3K9 appearing after birth. IUGR adult islets (bottom) are characterized by inactive chromatin with dimethylated H3K9 (Me) and extensive DNA methylation (dark grey circles) locking in the transcriptionally silent state of *Pdx1*.

DNA methylation of a CpG island in the promoter is a key mechanism for silencing gene expression. Most CpG islands remain unmethylated in normal cells; however, under some normal circumstances such as for imprinted genes and genes on the inactive X chromosome in females and for some disease processes such as cancer [49] and oxidative stress [50], CpG islands can become methylated de novo. This is particularly relevant to type 2 diabetes, as there are now substantial data that show that oxidative stress plays a significant role in the progression of beta cell deterioration [51–55]. Further, IUGR induces mitochondrial dysfunction in the beta cell leading to increased production of reactive oxygen species and oxidative stress [18]. It is not known why particular CpG islands are susceptible to aberrant methylation. A recent study by Feltus et al. [56] suggests that there is a 'sequence signature associated with aberrant methylation'. Of particular relevance to this study is their finding that *Pdx1* and a flanking gene, *Cdx2* (also encoding a homeobox protein), were 2 of only

15 genes (a total of 1,749 genes with CpG islands were examined) that were methylation susceptible under conditions of increased methylation induced by overexpression of DNMT1.

The molecular mechanism responsible for DNA methylation in IUGR islets is likely to involve H3K9 methylation. A number of studies have shown that methylation of H3K9 precedes DNA methylation [57, 58]. It has also been suggested that DNA methyltransferases may act only on chromatin that is methylated at lysine 9 on histone H3 (H3K9) [58]. Histone methyltransferases bind to the DNA methylases DNMT3A and DNMT3B, thereby initiating DNA methylation [59]. Surprisingly, we found that DNMT1 was associated with *Pdx1* prior to CpG methylation. Since DNMT1 can be recruited by interaction with a histone deacetylase such as HDAC1 [60], which is already associated with the *Pdx1* promoter in fetal IUGR islets, we suggest that this occurs in the IUGR islet prior to the alterations in histone methylation that only occur after birth (fig. 1). As H3K9 methylation occurs during postnatal life in IUGR islets, this would then allow recruitment of the de novo methyltransferase DNMT3A [59] (but not DNMT3B). Subsequent to onset of DNA methylation, DNMT1 would then be positioned to maintain the methylated state, locking in *Pdx1* silencing in adult IUGR islets (fig. 1).

Exendin-4 Treatment Reverses Aberrant Chromatin Remodeling

As described above, our earlier studies showed that treatment of IUGR newborn pups with exendin-4 for 6 days normalized Pdx1 expression. In preliminary studies, we determined that this normalization of Pdx1 expression was secondary to reversal of IUGR-induced chromatin remodeling at the proximal promoter. USF-1 binding to the proximal promoter of Pdx1 was restored by Ex-4 treatment of IUGR pups. This was accompanied by markedly enhanced histone acetylase activity. Increased histone acetylase activity was in turn linked to increased acetylation of histone H3 and histone H4 at the Pdx1 proximal promoter.

Conclusions

The permanent improvement in the long-term maintenance of beta cell mass induced by neonatal exendin-4 in the IUGR model suggests that there may be a unique opportunity to influence the development of adult-onset diabetes in humans by intervening during the prediabetic period in at-risk individuals. The newborn period in particular may represent a critical window in which therapies designed to enhance beta cell mass should be initiated. Refining the window for therapeutic intervention will be a subject of great interest in future studies.

References

1 Ogata ES, Bussey ME, Finley S: Altered gas exchange, limited glucose and branched chain amino acids, and hypoinsulinism retard fetal growth in the rat. Metabolism 1986;35:970–977.

2 Simmons RA, Gounis AS, Bangalore SA, Ogata ES: Intrauterine growth retardation: Fetal glucose transport is diminished in lung but spared in brain. Pediatr Res 1991;31:59–63.

3 Simmons RA, Flozak AS, Ogata ES: Glucose regulates Glut 1 function and expression in fetal rat lung and muscle in vitro. Endocrinology 1993;132:2312–2318.

4 Simmons RA, Templeton L: IUGR leads to apoptosis and decreased proliferation of β-cells of diabetic animals. Pediatr Res 1999;45:357A.

5 Simmons RA, Templeton L, Gertz S, Niu H: Intrauterine growth retardation leads to type II diabetes in adulthood in the rat. Diabetes 2001;50:2279–2286.

6 Gerich JE: The genetic basis of type 2 diabetes mellitus: impaired insulin secretion versus impaired insulin sensitivity. Endocr Rev 1998;19:491–503.

7 Kieffer TJ, Habener JF: The glucagon-like peptides. Endocr Rev 1999;20:876–913.

8 Scaglia L, Cahill CJ, Finegood DT, Bonner-Weir S: Apoptosis participates in the remodeling of the endocrine pancreas in the neonatal rat. Endocrinology 1997;138:1736–1741.

9 Bonner-Weir S: Perspective: postnatal pancreatic beta cell growth. Endocrinology 2000;141:1926–1929.

10 Stoffers DA, Desai BM, DeLeon DD, Simmons RA: Neonatal exendin-4 prevents the development of diabetes in the intrauterine growth retarded rat. Diabetes 2003;52:734–740.

11 Gedulin BR, Nikoulina SE, Smith PA, Gedulin G, Nielsen LL, Baron AD, Parkes DG, Young AA: Exenatide (Exendin-4) improves insulin sensitivity and β-cell mass in insulin-resistant obese fa/fa zucker rats independent of glycemia and body weight. Endocrinology 2005;146:2069–2076.

12 Young AA, Gedulin BR, Bhavzar S, Bodkin N, Jodka C, Hansen B, Denaro M: Glucose lowering and insulin-sensitizing actions of exendin-4: studies in obese diabetic (ob/ob, db/db) mice, diabetic fatty Zucker rats, and diabetic rhesus monkeys *(Macaca mulatta)*. Diabetes 1999;48:1026–1034.

13 Pospisilik JA, Stafford SG, Demuth HU, McIntosh CHS, Pederson RA: Long-term treatment with dipeptidyl peptidase IV inhibitor improves hepatic and peripheral insulin sensitivity in the VDF Zucker rat. Diabetes 2002;51:2677–2683.

14 Szayna M, Doyle ME, Betkey JA, Holloway HW, Spencer RGS, Greig NH, Egan JM: Exendin-4 decelerates food intake, weight gain, and fat deposition in Zucker rats. Endocrinology 2000;141:1936–1941.

15 Prigeon RL, Quddusi S, Paty B, D'Alessio DA: Suppression of glucose production by GLP-1 independent of islet hormones: a novel extrapancreatic effect. Am J Physiol Endocrinol Metab 2003;285: E701–E707.

16 Edlund H: Transcribing pancreas. Diabetes 1998;47: 1817–1823.

17 Brissova M, Shiota M, Nicholson WE, Gannon M, Knobel SM, Piston DW, Wright CV, Powers AC: Reduction in pancreatic transcription factor PDX-1 impairs glucose-stimulated insulin secretion. J Biol Chem 2002;277:11225–11232.

18 Simmons RA, Suponitsky-Kroyter I, Selak MA: Progressive accumulation of mitochondrial DNA mutations and decline in mitochondrial function lead to beta-cell failure. J Biol Chem 2005;280: 28785–28791.

19 Kaung HL: Growth dynamics of pancreatic islet cell populations during fetal and neonatal development of the rat. Dev Dyn 1994;200:163–175.

20 Dutta S, Bonner-Weir S, Montminy M, Wright CV: Regulatory factor linked to late-onset diabetes? Nature 1998;392:560.

21 Bannister AJ, Kouzarides T: Reversing histone methylation. Nature 2005;436:1103–1106.

22 Bernstein E, Allis CD: RNA meets chromatin. Genes Dev 2005;19:1635–1655.

23 Mellor J: The dynamics of chromatin remodeling at promoters. Mol Cell 2005;19:147–157.

24 Sproul D, Gilbert N, Bickmore WA: The role of chromatin structure in regulating the expression of clustered genes. Nat Rev Genet 2005;6:775–781.

25 Schubeler D, Lorincz MC, Cimbora DM, Telling A, Feng YQ, Bouhassira EE, Groudine M: Genomic targeting of methylated DNA: influence of methylation on transcription, replication, chromatin structure, and histone acetylation. Mol Cell Biol 2000;20: 9103–9112.

26 Costello JF, Fruhwald MC, Smiraglia DJ, Rush LJ, Robertson GP, Gao X, Wright FA, Feramisco JD, Peltomaki P, Lang JC, Schuller DE, Yu L, Bloomfield CD, Caligiuri MA, Yates A, Nishikawa R, Su Huang H, Petrelli NJ, Zhang X, O'Dorisio MS, Held WA, Cavenee WK, Plass C: Aberrant CpG-island methylation has non-random and tumour-type-specific patterns. Nat Genet 2000;24:132–138.

27 Galm O, Yoshikawa H, Esteller M, Osieka R, Herman JG: SOCS-1, a negative regulator of cytokine signaling, is frequently silenced by methylation in multiple myeloma. Blood 2003;101:2784–2788.
28 Yoshikawa H, Matsubara K, Qian GS, Jackson P, Groopman JD, Manning JE, Harris CC, Herman JG: SOCS-1, a negative regulator of the JAK/STAT pathway, is silenced by methylation in human hepatocellular carcinoma and shows growth-suppression activity. Nat Genet 2001;28:29–35.
29 He B, You L, Uematsu K, Zang K, Xu Z, Lee AY, Costello JF, McCormick F, Jablons DM: SOCS-3 is frequently silenced by hypermethylation and suppresses cell growth in human lung cancer. Proc Natl Acad Sci USA 2003;100:14133–14138.
30 Yoshida H, Broaddus R, Cheng W, Xie S, Naora H: Deregulation of the HOXA10 homeobox gene in endometrial carcinoma: role in epithelial-mesenchymal transition. Cancer Res 2006;66:889–897.
31 Shiraishi M, Sekiguchi A, Oates AJ, Terry MJ Miyamoto Y: HOX gene clusters are hotspots of de novo methylation in CpG islands of human lung adenocarcinomas. Oncogene 2002;21:3659–3663.
32 Kim JY, Siegmund KD, Tavare S, Shibata D: Age-related human small intestine methylation: evidence for stem cell niches. BMC Med 2005;3:10–15.
33 So K, Tamura G, Honda T, Homma N, Waki T, Togawa N, Nishizuka S, Motoyama T: Multiple tumor suppressor genes are increasingly methylated with age in non-neoplastic gastric epithelia. Cancer Sci 2006;97:1155–1158.
34 Takahashi T, Shigematsu H, Shivapurkar N, Reddy J, Zheng Y, Feng Z, Suzuki M, Nomura M, Augustus M, Yin J, Meltzer SJ, Gazdar A: Aberrant promoter methylation of multiple genes during multistep pathogenesis of colorectal cancers. Int J Cancer 2006;118: 924–931.
35 Feltus FA, Lee EK, Costello JF, Plass C, Vertino PM: Predicting aberrant CpG island methylation. Proc Natl Acad Sci USA 2003;100:12253–12258.
36 MacLennan NK, James SJ, Melnyk S, Piroozi A, Jernigan S, Hsu JL, Janke SM, Pham TD, Lane RH: Uteroplacental insufficiency alters DNA methylation, one-carbon metabolism, and histone acetylation in IUGR rats. Physiol Genomics 2004;18:43–50.
37 Pham TD, MacLennan NK, Chiu CT, Laksana GS, Hsu JL, Lane RH: Uteroplacental insufficiency increases apoptosis and alters p53 gene methylation in the full-term IUGR rat kidney. Am J Physiol Regul Integr Comp Physiol 2003;285:R962–R970.
38 Fu Q, McKnight RA, Yu X, Wang L, Callaway CW, Lane RH: Uteroplacental insufficiency induces site-specific changes in histone H3 covalent modifications and affects DNA-histone H3 positioning in day 0 IUGR rat liver. Physiol Genomics 2004;20:108–116.
39 Park JH, Stoffers DA, Nicholls RD, Simmons RA: Development of type 2 diabetes following intrauterine growth retardation in rats is associated with progressive epigenetic silencing of Pdx1. J Clin Invest 2008;118:2316–2324.
40 Qian J, Kaytor EN, Towle HC, Olson LK: Upstream stimulatory factor regulates Pdx-1 gene expression in differentiated pancreatic β-cells. Biochem J 1999; 341:315–322.
41 Sharma S, Leonard J, Lee S, Chapman HD, Leiter EH, Montminy MR: Pancreatic islet expression of the homeobox factor STF-1 *(Pdx-1)* relies on an E-box motif that binds USF. J Biol Chem 1996;71: 2294–2299.
42 Shi Y, Whetstine JR: Dynamic regulation of histone lysine methylation by demethylases. Mol. Cell 2007; 25:1–14.
43 Forneris F, Binda B, Dall'Aglio A, Fraaije MW, Battaglioli E, Mattevi A: A highly specific mechanism of histone H3-K4 recognition by histone demethylase LSD1. J Biol Chem 2006;281:35289–35295.
44 Klose RJ, Yan Q, Tothova Z, Yamane K, Erdjument-Bromage H, Tempst P, Gilliland DG, Zhang Y, Kaelin WGJ: The retinoblastoma binding protein RBP2 is an H3K4 demethylase. Cell 2007;128:889–900.
45 Trewick SC, Henshaw TF, Hausinger RP, Lindahl T, Sedgwick B: Oxidative demethylation by Escherichia coli AlkB directly reverts DNA base damage. Nature 2002;419:174–178.
46 Cloos PA, Christensen J, Agger K, Maiolica A, Rappsilber J, Antal T, Hansen KH, Helin K: The putative oncogene GASC1 demethylates tri- and dimethylated lysine 9 on histone H3. Nature 2006; 42:307–311.
47 Litt MD, Simpson M, Gaszner M, Allis CD, Felsenfeld G: Correlation between histone lysine methylation and developmental changes at the chicken beta-globin locus. Science 2001;293:2453–2455.
48 Noma K, Grewal SI: Histone H3 lysine 4 methylation is mediated by Set and promotes maintenance of active chromatin states in fission yeast. Proc Natl Acad Sci USA 2002;99:16438–16445.
49 He B, You L, Uematsu K, Zang K, Xu Z, Lee AY, Costello JF, McCormick F, Jablons DM: SOCS-3 is frequently silenced by hypermethylation and suppresses cell growth in human lung cancer. Proc Natl Acad Sci USA 2003;100:14133–14138.
50 Cerda S, Weitzman SA: Influence of oxygen radical injury on DNA methylation. Mutat Res 1997,386: 141–152.

51 Ihara Y, Toyokuni S, Uchida K, Odaka H, Tanaka T, Ikeda H, Hiai H, Seino Y, Yamada Y: Hyperglycemia causes oxidative stress in pancreatic beta-cells of GK rats, a model of type 2 diabetes. Diabetes 1999;48:927–932.
52 Silva JP, Kohler M, Graff C, Oldfors A, Magnuson MA, Berggren PO, Larsson NG: Impaired insulin secretion and β-cell loss in tissue specific knockout mice with mitochondrial diabetes. Nat Genet 2000; 26:336–340.
53 Kaneto H, Xu G, Song KH, Suzuma K, Bonner-Weir S, Sharma A, Weir GC: Activation of the hexosamine pathway leads to deterioration of pancreatic beta-cell function through the induction of oxidative stress. J Biol Chem 2001;276:31099–31104.
54 Sakuraba H, Mizukami H, Yagihashi N, Wada R, Hanyu C, Yagihashi S: Reduced beta-cell mass and expression of oxidative stress-related DNA damage in the islet of Japanese Type II diabetic patients. Diabetologia 2002;45:85–96.
55 Sakai K, Matsumoto K, Nishikawa T, Suefuji M, Nakamaru K, Hirashima Y, Kawashima J, Shirotani T, Ichinose I, Brownlee M, et al: Mitochondrial reactive oxygen species reduce insulin secretion by pancreatic β-cells. Biochem Biophys Res Commun 2003;300:216–222.
56 Feltus FA, Lee EK, Costello JF, Plass C, Vertino PM: Predicting aberrant CpG island methylation. Proc Natl Acad Sci USA 2003;100:12253–12258.
57 Bachman KE, Park BH, Rhee I, Rajagopalan H, Herman JG, Baylin SB, Kinzler KW, Vogelstein B: Histone modifications and silencing prior to DNA methylation of a tumor suppressor gene. Cancer Cell 2003;3:89–95.
58 Kouzarides T: Histone methylation in transcriptional control. Curr Opin Genet Dev 2002;12:198–209.
59 Li H, Rauch T, Chen ZX, Szabo PE, Riggs AD, Pfeifer GP: The histone methyltransferase SETDB1 and the DNA methyltransferase DNMT3A interact directly and localize to promoters silenced in cancer cells. J Biol Chem 2006;281:19489–19500.
60 Fuks F, Burgers WA, Brehm A, Hughes-Davies L, Kouzarides T: DNA methyltransferase Dnmt1 associates with histone deacetylase activity. Nat Genet 2000;24:88–91.

Dr. Rebecca A. Simmons
Department of Pediatrics, Children's Hospital Philadelphia
BRB II/III, Rm 1308, University of Pennsylvania
421 Curie Blvd, Philadelphia, PA 19104 (USA)
Tel. +1 215 746 5139, Fax +1 215 573 7627, E-Mail rsimmons@mail.med.upenn.edu

Newnham JP, Ross MG (eds): Early Life Origins of Human Health and Disease.
Basel, Karger, 2009, pp 184–192

Prenatal Risk Factors for Breast Cancer and Clues to the Underlying Biological Mechanisms from Animal and Human Studies

Rebecca Troisi[a,b] · Isabel dos Santos Silva[d] · Retha Newbold[c]

[a]Division of Cancer Epidemiology and Genetics, National Cancer Institute, National Institutes of Health, Department of Health and Human Services, Bethesda, Md., [b]Department of Community and Family Medicine, Dartmouth Medical School, Hanover, N.H., [c]Developmental Endocrinology and Endocrine Disruptor Section, Laboratory of Molecular Toxicology, National Institute of Environmental Health Sciences, National Institutes of Health, Department of Health and Human Services, Research Triangle Park, N.C., USA; [d]Department of Epidemiology and Population Health, London School of Hygiene and Tropical Medicine, London, UK

Epidemiologic and animal evidence, in addition to data from migrant studies, suggests that breast carcinogenesis may have its origins in part during the prenatal period. The in utero hypothesis, put forward by Trichopoulos [1] in 1990, described a novel paradigm for breast cancer epidemiology – that the developing breast is influenced by the fetal environment, particularly variations in hormone concentrations which could mediate subsequent breast cancer development. In response, a growing body of epidemiologic, animal and clinical studies assessing the offspring's cancer risk with maternal and perinatal characteristics and exploring potential underlying biological mechanisms has accumulated.

Opportunities for direct investigations of in utero influences on adult cancer risk have been limited; thus, an indirect approach has evolved with investigation of pregnancy characteristics as proxies for biochemical and molecular exposures and cancer risk in epidemiologic studies (fig. 1). The biological correlates of these risk factors are then described and inferences made regarding biological mechanisms of carcinogenesis. Animal studies have both stimulated, as in the case of diethylstilbestrol (DES) [2], and followed up on these leads, providing additional data to assess biological hypotheses.

In this chapter, we summarize the state of the literature on prenatal epidemiologic risk factors for offspring's cancer risk, the animal data on DES and their implications for the in utero estrogen hypothesis, and the human biological data that address this hypothesis as well.

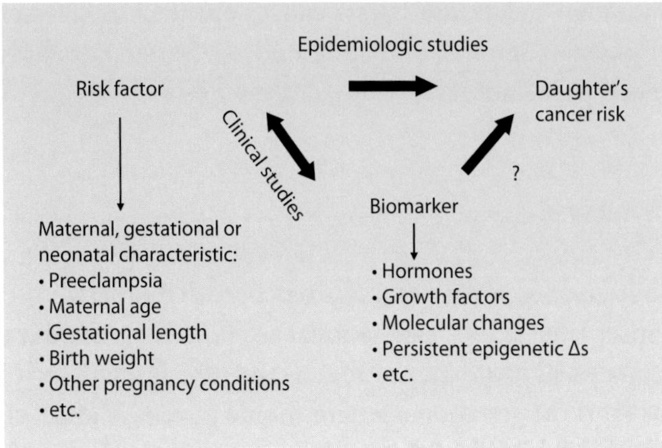

Fig. 1. Indirect approach for investigating in utero influences on adult cancer risk.

Prenatal Risk Factors for Breast Cancer – The Epidemiological Evidence

A large number of epidemiological studies have attempted to investigate the prenatal origins of breast cancer by relying on indirect markers of the intrauterine environment. The most investigated markers have been birth size, gestational age, twinship and maternal preeclampsia.

Size at Birth

A recent pooled analysis [3] has examined the relationship between birthweight, the widest available measure of birth size, and subsequent breast cancer risk. Most studies based on birth records provide effect estimates consistent with a modest increase in breast cancer risk with increasing birthweight (pooled RR per 0.5 kg increase in birthweight: 6% (95% CI:2–9%), and further adjustment for other maternal and perinatal characteristics and for adult risk factors does not affect the magnitude of this association. In contrast, studies of birthweight based on reports in adulthood by the women themselves or by their mothers showed no association between birthweight and breast cancer risk, probably reflecting greater misclassification of birthweight than in studies based on birth records.

Birthweight reflects both linear growth and adiposity and either of these (or other unmeasured factors correlated with birthweight) may explain its positive association with breast cancer risk. Birth length, a marker of linear growth, shows a modest positive association with breast cancer risk [3] in studies based on birth records pooled RR per 2 cm increase in length adjusting for weight: 9% (95% CI: 0–19%). There was no association between ponderal index (a measure of adiposity) and risk. The association between birth size and breast cancer risk may be mediated by postnatal growth, as birthweight predicts early growth and development, and age at menarche and adult height are breast cancer risk factors. However, adjustment for measures of postnatal growth and adult body size attenuates only slightly the birth size-breast cancer association [3].

Gestational Age

Shorter pregnancy gestation has been hypothesized to be protective against breast cancer later in life as gestational age represents the duration of exposure of the breast to maternal pregnancy estrogens and other growth factors [4]. An alternate hypothesis is that short gestational length should adversely affect risk because postnatal estrogen secretion is enhanced in girls born prematurely [5]. Several studies have examined the relationship between gestational age and breast cancer risk [6], but their findings are largely inconsistent. Difficulties in obtaining accurate measures of gestational age, small sample sizes, and the lack of adjustment for birth size, other perinatal variables and adult life risk factors limit the interpretation of findings from these studies.

Twinship

Most epidemiological studies, but not all, have found an elevated risk of breast cancer in twin daughters compared to singletons [6]. Results of studies comparing risk in dizygotic (DZ) with monozygotic (MZ) twins or singletons are less consistent, partly due to difficulties in establishing zygosity, with some studies being consistent with the hypothesis of a higher risk among DZ twins, but not others. It is possible that risk is higher in women with a twin brother, as a proxy for being DZ, than in those with a twin sister, which is a mixture of MZ and DZ twins. Furthermore, having a male co-twin may have an effect beyond that of zygosity, but the association with co-twin's gender is not consistently demonstrated in all studies.

Maternal Preeclampsia

Preeclampsia, a pregnancy complication characterized by hypertension and proteinuria, has been hypothesized to reduce breast cancer risk in women born of these pregnancies [7]. The earliest study to address this issue reported a nonsignificant increase in the risk of breast cancer with maternal preeclampsia, but was based on only 4 exposed cases. In contrast, a Swedish population-based case-control study reported a strong protective effect of maternal preeclampsia (RR = 0.41; 95% CI = 0.22–0.79; based on 14 exposed cases) which persisted after adjustment for other maternal and perinatal variables. Evidence in favor of a protective effect of preeclamptic pregnancies was also provided by data from three other studies, all based on small numbers of exposed cases. Thus, the limited available data appear consistent with an inverse association between maternal preeclampsia and breast cancer risk in women born from such pregnancies, but confirmation is warranted in larger studies with more detailed information on preeclampsia diagnosis and data on potential confounders and effect modifiers.

Animal Studies Using Diethylstilbestrol as a Model for in utero Estrogen Exposure

Accumulating evidence from animal studies support the hypothesis that breast (mammary) cancer has an origin in prenatal life. This is a theory that is receiving renewed attention [8–10]. The developmental period is considered particularly susceptible to perturbation by environmental insult because of the high rates of cell proliferation and extensive differentiation that occur, thus offering multiple opportunities for mutagenic and epigenetic alterations. Further, physiologic protective barriers such as the blood-brain barrier are not complete, and the metabolizing and eliminating capabilities of the developing organism are not fully developed in utero. One example of the sensitivity of the developing fetus can be seen by the carcinogenic effects of prenatal exposure to ionizing radiation [11]. In addition, ample evidence links prenatal exposure to DES, a potent estrogenic chemical, with breast and reproductive tract cancers later in life. In fact, DES holds the dubious position of being the first definitely established transplacental chemical carcinogen in humans; DES was shown to cross the placenta and to induce a direct effect on the developing fetus, but these effects were not identified until later in life, long after exposure elapsed [11, 12].

Briefly, DES, a potent synthetic estrogen, was prescribed during the late 1940s-1970s initially to women with high risk pregnancies with the mistaken belief that it could prevent miscarriage and other complications of pregnancy. However, DES exposure was shown to cause a low but significant increased risk of vaginal cancer in the daughters of these women, and an increased risk of reproductive problems including infertility/subfertility in both their sons and daughters [for review, see 12, 13]. Recent epidemiological data also show a significantly increased incidence of breast cancer in DES-exposed daughters as they age [14]. Although DES is no longer used clinically to prevent miscarriage, many uncertainties remain such as: (1) will the incidence of cancer and other problems continue to rise as the DES-exposed population ages, (2) will prenatal DES exposure cause abnormal responses to other hormone therapies later in life such as oral contraceptives or hormone replacement therapies, and (3) can susceptibility to tumors like breast cancer be passed to the next generation?

To help answer these questions, along with the need to identify mechanisms involved in these DES-induced carcinogenic effects, numerous animal models have been developed through the years to study the effects of estrogens on mammary

Table 1. Comparative effects of prenatal exposure to DES in mice and humans

	Male offspring	Female offspring
Reproductive tract dysfunction	subfertility/infertility decreased sperm counts	subfertility/infertility poor reproductive outcome altered estrous (menstrual) cycles
Structural malformations	microphallus and hypospadias retained hypoplastic testes retained mullerian remnants (anatomical feminization)	oviduct, uterus, cervix and vagina paraovarian cysts of mesonephric origin retained mesonephric remnants
Cellular abnormalities	tumors – testis – retained mullerian ducts – prostate and seminal vesicle epididymal cysts inflammation	lesions (proliferative) – oviduct – uterus/cervix uterine fibroids vaginal adenosis vaginal adenocarcinoma breast (mammary) cancer

Modified from Newbold [2].

gland and reproductive tract differentiation. The murine model using outbred mice has been most successful in duplicating and predicting adverse effects seen in humans with similar DES exposure [summarized in 2]. Table 1 is a compilation of the reported effects in mice and humans. In particular, there is an association of breast cancer with prenatal DES treatment in both species. Consistent with this observation, mammary carcinogenesis has been extensively documented in numerous other experimental animal models following prenatal DES administration [15–17]. Furthermore, the extensive list of detrimental effects across organ systems in both species supports the idea that chemicals with estrogenic activity can perturb multiple developing systems especially if exposure occurs during critical windows of differentiation, and that this perturbation can result in carcinogenesis as one outcome.

Subsequent to the myriad of DES reports, in the early 1990s, there was rising concern that exposure to other chemicals in the environment with hormone-like activity (estrogenic as well as antiestrogenic, androgenic, antiandrogenic, progesterone-like, thyroid-like, etc.) may be contributing to increases in testicular cancer, undescended testes, and malformations of the male genital tract observed over the last 50 years [18]. This correlation was quickly extended to include the increased incidence of breast cancer during the same time interval [19]. Again, the prenatal and neonatal periods of development were considered important risk factors for the development of cancer. Since then, significant progress has been made in establishing that the perinatal development period is extremely sensitive to perturbation by chemical with hormone-like activity; these chemicals are now collectively called 'endocrine disruptors.' Exposure to low, environmentally relevant doses of xenoestrogens induces morphological and functional alterations in the male [20] and female genital tracts [2] and the mammary gland [8, 9, 21, 22]. Using experimental animal models, alterations have been shown in the mammary gland that include early maturation and altered morphogenesis. Specifically, there was a decrease in the migration of the ductal epithelium into the stromal compartment, but also a permanent increase in the number of terminal ducts and terminal end buds. Such changes have been associated with enhanced susceptibility to carcinogenesis in both rodents and humans. The underlying mechanisms suggest that many of these morphological alterations may have a molecular basis.

The *homeobox* (HOX) and *wingless* (Wnt) families of genes have been identified as hormone-responsive candidates that relay information on tissue patterning within the developing mammary gland and reproductive tract, particularly regarding the relationship between the epithelial and stromal compartments, which is critical to normal development. For example, MSX 1, MSX 2, and Wnt 10b are expressed during prenatal mammary gland morphogenesis [23]. The expression of some of these Hox and Wnt genes was shown to change during the murine estrous cycle as the circulating levels of steroid hormones changed and that they were also downregulated following ovariectomy. Most importantly, prenatal exposure to DES was shown to alter expression of Wnt 7a and Hoxa-10 during uterine morphogenesis in the mouse, thus establishing an important correlation between chemical exposure and the ensuing developmental abnormalities associated with endocrine disruption [23, 24]. This link was made even more credible when the phenotype of mice carrying specific HOX and Wnt null mutations was observed to be similar to that of human and mouse 'DES' daughters. Wnt genes are associated with cellular responses such as cell proliferation, apoptosis, and cell-cell communication (through the B-catenin/E-cadherin complex). Changing patterns of expression due to environmental chemical exposure can thus provide insight into how these chemicals influence cell fate determination and tissue morphogenesis in hormone-sensitive organs.

In summary, the data from experimental animal studies suggesting that developmental exposures are risk factors for adult breast cancer are far more extensive and convincing than current epidemiological data. However, many human studies are designed to measure hormone or chemical exposure levels only at the time of detection of breast cancer. Growing evidence from animal studies suggests that the critical window of exposure occurs much earlier and the association will have to be determined at the time of exposure not later at the time of tumor detection.

The Consistency of Human Biological Evidence with Current Hypotheses for Hormonal Mechanisms Underlying the Associations of Prenatal Risk Factors for Breast Cancer

The most seductive hypotheses for underlying biological mechanisms are those that attempt to explain all or virtually all epidemiological, prenatal risk factors. The most prominent hypothesis has been that fetal estrogen exposure is directly associated with subsequent breast cancer risk [1]. In this section, we present data from human studies characterizing the biological correlates of prenatal breast cancer risk factors to determine whether the evidence is consistent with this hypothesis.

If the effects of high birthweight and DZ twinning are mediated through elevations in pregnancy estrogen concentrations, and in the case of preeclampsia, a reduction, then estrogen concentrations should be positively associated with birthweight and DZ twinning (or inversely associated with preeclampsia) [reviewed in detail in 10]. Maternal estriol concentrations are clearly elevated in the circulation of high birthweight pregnancies, but whether birth size is associated with maternal estradiol is unclear. Furthermore, there is little evidence that birthweight is associated with estrogens or androgens in cord blood. Overall, the data imply that birthweight is positively associated with maternal but not fetal estrogen concentrations. There is a paucity of sufficiently large, well-designed studies on hormone levels in twin pregnancies. While one study showed over 50% greater mid-pregnancy serum estradiol concentrations in mothers of twins than those of singletons, it was based on only 11 twin pregnancies. Levels of estrogens and other hormonal or endocrine factors have not been studied by number of placentae, and there are no data on estrogen concentrations in the cord blood of twin pregnancies. Maternal urinary estriol excretion declines late in preeclamptic pregnancies; however, estrogens in the circulation near delivery do not appear to be lower in preeclampsia compared with uncomplicated pregnancy. Furthermore, the limited data are not consistent with lower umbilical cord blood estriol, estradiol or estrone concentrations in preeclampsia. Clearly, additional data are warranted on maternal and cord estrogen concentrations in preeclamptic and uncomplicated pregnancies, as well as for other established prenatal breast cancer risk factors. A more direct test of the prenatal estrogen hypothesis derives from results of the National Cancer Institute's DES cohort study which found no excess in breast cancer risk in exposed women under 40 years of age, though recent data show a steadily increasing risk from age 40 through 50 years and older [14].

Thus, the available data are not consistent with the prevailing hypothesis that elevated in utero endogenous estrogen exposure is a unifying hypothesis that explains most identified prenatal risk factors for breast cancer risk in daughters, though recent findings for pharmacologic exposure to estrogen during pregnancy show an excess relative risk of breast cancer in women over the age of 40 years. Whether this is due to estrogenic activity or other carcinogenic aspects of DES, or alterations in endogenous hormones resulting from high pharmacologic estrogen doses remains to be determined.

There are many metabolic changes that accompany pregnancy including alterations in factors that have been investigated for their involvement in breast cancer risk. Insulin-like growth factors, α-fetoprotein, hCG, immunological factors, pro- and antiangiogenic factors and numerous other biological compounds should be investigated. To take just one example, elevated exposure to another class of hormones, the androgens, might confer long-term protection for breast cancer in the daughter by antagonizing the effect of estrogens on ductal development in the fetal breast or limiting the initial population of breast stem cells [10]. Elevated androgen concentrations have been observed in preeclamptic pregnancies and in Asian and African-American compared with Caucasian pregnancies. These observations would be consistent with a protective effect of prenatal androgens on breast cancer risk but the data, particularly for cord concentrations, are sparse and there is no evidence of a positive association of androgen concentrations with birthweight or twinning.

It is premature at this point in our understanding of the biology of pregnancy characteristics related to subsequent breast cancer risk in the daughters to focus exclusively on one or two of these candidate mechanisms. Instead, we should seek to uncover the plausible range of hypotheses, based on a comprehensive characterization of the biochemical and molecular consequences of these risk factors.

References

1 Trichopoulos D: Hypothesis: does breast cancer originate in utero? Lancet 1990;335:939–940.
2 Newbold RR: Lessons learned from perinatal exposure to diethylstilbestrol. Toxicol Appl Pharmacol 2004;199:142–150.
3 dos Santos Silva I, De Stavola B, McCormack V, Collaborative Group on Pre-Natal Risk Factors and Subsequent Risk of Breast Cancer: Birth size and breast cancer risk: re-analysis of individual participant data from 32 studies. PLoS Med 2008,5:e193. doi:10.1371/journal.pmed.0050193.
4 McCormack VA, dos Santos Silva I, De Stavola BL, Mohsen R, Leon DA, Lithell HO: Foetal growth and subsequent risk of breast cancer: results from long-term follow-up of Swedish cohort. BMJ 2003;326: 248–251.
5 Ekbom A, Erlandsson G, Hsieh C-C, Trichopoulos D, Adami H-O, Cnattingius S: Risk of breast cancer in prematurely born women. J Natl Cancer Inst 2000;92:840–841.
6 Potischman N, Troisi R, Vatten L: The life course approach to cancer epidemiology; in Kuh D, Ben-Shlomo Y (eds): A Life Course Approach to Chronic Disease Epidemiology, ed 2. New York, Oxford University Press, 2004.
7 Innes KE, Byers TE: Preeclampsia and breast cancer risk. Epidemiology 1999;10:722–732.
8 Birnbaum LS, Fenton SE: Cancer and developmental exposure to endocrine disruptors. Environ Health Perspect 2003;111:389–394.
9 Hilakivi-Clarke L, de Assis S: Fetal origins of breast cancer. Trends Endocrinol Metab 2006;17:340–348.
10 Troisi R, Potischman N, Hoover RN: Exploring the underlying hormonal mechanisms of prenatal risk factors for breast cancer: a review and commentary. Cancer Epidemiol Biomarkers Prev 2007;16:1700–1712.
11 Tomatis L: Overview of perinatal and multigeneration carcinogenesis. IARC Sci Publ 1989;96:1–15.
12 Herbst AL, Bern HA: Developmental Effects of Diethylstilbestrol (DES) in Pregnancy. New York, Thieme-Stratton, 1981.
13 NIH Workshop Update on Long-Term Effects of Exposure to Diethylstilbestrol (DES): NIH Publication 00-4722, DHHS, 1999.
14 Palmer JR, Hatch EE, Rosenberg CL, Hartge P, Kaufman RH, Titus-Ernstoff L, Noller KL, Herbst AL, Rao RS, Troisi R, Colton T, Hoover RN: Risk of breast cancer in women exposed to diethylstilbestrol in utero: preliminary results (United States). Cancer Cause Control 2002;13:753–758.

15 Rothschild TC, Boylan ES, Calhoon RE, Vonderhaar BK: Transplacental effects of diethylstilbestrol on mammary development and tumorigenesis in female ACI rats. Cancer Res 1987;47:4508–4516.
16 Bern HA, Edery M, Mills KT, Kohrman AF, Mori T, Larson L: Long-term alterations in histology and steroid receptor levels of the genital tract and mammary gland following neonatal exposure of female BALB/cCrgl mice to various doses of diethylstilbestrol. Cancer Res 1987;47:4165–4172.
17 Mori T, Bern HA, Mills KT, Young PN: Long-term effects of neonatal steroid exposure on mammary gland development and tumorigenesis in mice. J Natl Cancer Inst 1976;57:1057–1062.
18 Sharpe RM, Skakkebaek NE: Are oestrogens involved in falling sperm counts and disorders of the male reproductive tract? Lancet 1993;341:1392–1395.
19 Davis DL, Bradlow HL, Wolff M, Woodruff T, Hoel DG, Anton-Culver H: Medical hypothesis: xenoestrogens as preventable causes of breast cancer. Environ Health Perspect 1993;101:372–377.
20 vom Saal F, Timms BG, Montano MM, Palanza P, Thayer KA, Nagel SC, Ganjam VK, Parmigiani S, Welshons WV: Prostate enlargement in mice due to fetal exposure to low doses of estradiol or diethylstilbestrol and opposite effects at high doses. Proc Natl Acad Sci USA 1997;2056–2061.
21 Markey CM, Luque EH, Munoz De Toro M, Sonnenschein C, Soto AM: In utero exposure to bisphenol A alters the development and tissue organization of the mouse mammary gland. Biol Reprod 2001;65:1215–1223.
22 Padilla-Banks E, Jefferson WN, Newbold RR: Neonatal exposure to the phytoestrogen genistein alters mammary gland growth and developmental programming of hormone receptor levels. Endocrinology 2006;147:4871–4882.
23 Miller C, Degenhardt K, Sassoon DA: Fetal exposure to DES results in de-regulation of Wnt7a during uterine morphogenesis. Nat Genet 1998;20:228–230.
24 Kitajewski J, Sassoon D: The emergence of molecular gynecology: homeobox and Wnt genes in the female reproductive tract. Bioessays 2000;22:902–910.

Dr. Rebecca Troisi
Dartmouth-Hitchcock Medical Center
Room 854, 7297 Rubin Building
One Medical Center Drive, Lebanon, NH 03756 (USA)
Tel. +1 603 653 9024, Fax +1 603 653 9093, E-Mail troisir@mail.nih.gov

Newnham JP, Ross MG (eds): Early Life Origins of Human Health and Disease.
Basel, Karger, 2009, pp 193–204

The Offspring of Women with Severe Mental Disorder

V.A. Morgan[a,b] · J. McGrath[d] · C.M. Hultman[e] · S. Zubrick[c] · C. Bower[c] · M. Croft[a] · G. Valuri[a] · A.V. Jablensky[a,b]

[a]Neuropsychiatric Epidemiology Research Unit, [b]Centre for Clinical Research in Neuropsychiatry, School of Psychiatry and Clinical Neurosciences, [c]Telethon Institute for Child Health Research, Centre for Child Health Research, The University of Western Australia, Perth, W.A., [d]Queensland Centre for Mental Health Research and Queensland Brain Institute, University of Queensland, Brisbane, QLD, Australia; [e]Department of Medical Epidemiology and Biostatistics, Karolinska Institutet, Stockholm, Sweden

Schizophrenia is a severely disabling disorder with a lifetime risk of about 1%. The aetiology of schizophrenia is complex. Evidence from twin, adoptive and family studies indicates that the disorder involves a significant genetic contribution: children of women with schizophrenia have an 8- to 10-fold higher risk of developing the disorder compared with the general population [1], with the risk in a monozygotic co-twin increased to about 50-fold [2]. The pattern of inheritance in schizophrenia is not a simple Mendelian one. It is likely that multiple genes of small to moderate effect contribute to the disorder, and that environmental risk factors interact with this genetic susceptibility. However, the precise nature of the genetic and environmental risk factors remains unclear.

It has long been suspected that disruptions to early brain development may be associated with an increased risk of adult-onset schizophrenia [3]. In the 1980s, these speculations became more clearly articulated under the generic label of the 'neurodevelopmental hypothesis' of schizophrenia [4]. It was proposed that critical circuits in the brain were affected in early development by a disease process, with full-blown consequences evident many years later in adolescence or early adulthood as schizophrenia [5]. The original theory is open as to whether the 'disease process' affecting neuronal development is genetically or environmentally determined. There are a number of indicators of neurodevelopmental disruption in schizophrenia including: (a) an increased risk of schizophrenia following obstetric complications at birth with good evidence that the risk is increased 2- to 7-fold [6]; (b) a history of neurointegrative defect in children who later develop schizophrenia [7], and (c) an increased frequency of minor physical anomalies [8] and neurological soft signs [9] in affected persons, indicative of a neurodevelopmental disturbance. In light of this, there continues to be considerable interest in understanding schizophrenia as a disorder of neurodevelopment, with a strong emphasis on the role of obstetric complications. In particular, pathological mechanisms that may interfere with neuronal migration during fetal development have been targeted including adverse hypoxic-ischaemic events [6], those affecting immune responses [10] and maternal undernutrition [11]. In more recent years, this line of schizophrenia research has converged with research under the rubric of developmental origins of health and disease [12].

The three studies that follow illustrate some of the work currently in progress in the area of neurodevelopment and schizophrenia. Studies 1 and 2 are record linkage studies, using population-based registers in Western Australia and Sweden, respectively, to follow up outcomes for the children of mothers with schizophrenia. Study 1 has a focus on the relationship between genetic and environmental (obstetric) risk factors and adverse outcomes for these children. The data presented here describe a range of adverse obstetric and offspring outcomes in the perinatal period, including the risk of birth defects. Study 2 also looks at outcomes for the children of mothers with schizophrenia, and is specifically interested in fetal growth and neonatal death and the impact of maternal risk factors such as exposure to smoking on offspring outcomes. In addition, it broadens its scope to include fathers with schizophrenia in the analysis. Finally, study 3 is concerned with schizophrenia as an outcome, rather than an exposure. It takes a neurobiological approach in order to understand better the role of one candidate risk factor for schizophrenia, prenatal exposure to low maternal vitamin D.

Study 1. The Neurodevelopmental Hypothesis and Schizophrenia: Evidence from the Western Australian Study of Women with Severe Mental Illness

The first study comes from a team of researchers in the Neuropsychiatric Epidemiology Research Unit within the School of Psychiatry and Clinical Neurosciences at the University of Western Australia, working under the direction of Prof. Assen Jablensky. This study is designed to untangle genetic and environmental contributions to the risk for schizophrenia and other adverse outcomes in the children of mothers with schizophrenia and other severe mental illness using Western Australian whole-of-population health databases. Records for 79,599 women on the Western Australian psychiatric case register were cross-linked to 308,022 birth records on the midwives database. Women with psychosis who had given birth in Western Australia between 1980 and 1992 were identified. Comparison mothers were randomly selected from mothers with no record of psychiatric illness. The study database comprised a total of 3,662 mothers and 6,303 children: 382 mothers with schizophrenia (618 children); 763 mothers with bipolar disorder (1,301 children); 686 mothers with unipolar major depression (1,255 children), and 1,831 comparison mothers (3,129 children).

Table 1. Selected obstetric complications (odds ratio, 95% confidence interval)

	Maternal diagnosis		
	Schizophrenia vs. comparison	Bipolar vs. comparison	Unipolar vs. comparison
Maternal socio-demographic characteristics			
Age at delivery ≤19	1.5 (1.0–2.2)	NS	NS
Age at delivery ≥35	2.0 (1.4–2.7)	NS	1.3 (1.0–1.7)
Single/widowed/divorced	4.3 (3.4–5.5)	2.1 (1.7–2.6)	1.5 (1.2–2.0)
Partner unemployed/pensioner	6.0 (2.2–16.9)	NS	NS
Aboriginal	2.7 (1.7–4.2)	NS	NS
Pregnancy complications			
Composite scale score[1]: pregnancy complications	1.4 (1.2–1.8)	1.2 (1.1–1.4)	1.2 (1.0–1.4)
Specific pregnancy complications			
Placenta praevia	NS	2.0 (1.1–3.7)	NS
Abruption of the placenta	2.8 (1.3–5.7)	NS	NS
Other antepartum haemorrhage	1.6 (1.0–2.7)	1.7 (1.2–2.4)	NS
Side effects of street drugs	3.8 (1.1–12.8)	3.9 (1.5–10.6)	NS
Labour/delivery complications (maternal)			
Composite scale score[1]: labour and delivery complications	NS	NS	NS
Neonatal complications			
Composite scale score[1]: neonatal complications	1.3 (1.0–1.5)	NS	NS
Specific neonatal complications			
Fetal distress	1.4 (1.1–1.8)	NS	NS
Narcotic antagonist used	1.9 (1.2–2.9)	NS	1.6 (1.1–2.3)
Percent of expected birthweight <10th percentile	1.4 (1.0–1.9)	NS	NS
Birth defects			
Any birth defects	NS	NS	NS
Cardiovascular defects	2.6 (1.2–5.5)	NS	NS
Other defects (mainly minor physical anomalies)	2.2 (1.1–4.5)	NS	NS

NS = Not significant.

[1] McNeil-Sjöström Scale for measuring obstetric complications. Severity level 4 [14, 15].

Fathers were identified using birth registrations. Full psychiatric histories for mothers, fathers and children were extracted, and data collected on obstetric complications and other childhood morbidities [for a more detailed description of the study design, see Jablensky et al. 13]. The study aims were to: (a) determine the frequency and distribution of obstetric complications in women with schizophrenia compared with a non-psychiatric comparison group of mothers; (b) explore the spectrum of outcomes for the children born to women with schizophrenia; (c) assess specificity of findings to maternal schizophrenia compared with maternal bipolar disorder and unipolar major depression, and (d) examine the relationship between familial psychiatric status, obstetric complications and mental health outcomes for children (in work in progress). The main study findings to date are summarised in table 1.

The Western Australian researchers found that mothers in all three diagnostic groups were more likely than comparison women to be single mothers, and to experience socioeconomic disadvantage relative to the general population. In addition, women with schizophrenia were characterized by a significant excess of either very young (age ≤19) or older (age ≥35) mothers. They were also more likely to have partners who were unemployed or disabled.

With respect to obstetric complications, this research team found that crude odds ratios (ORs) calculated using the McNeil-Sjöström summated scores indicated that mothers in all three diagnostic groups were significantly more likely to have complications during the pregnancy period. The risk of labour and delivery complications was not increased for any diagnostic group, while the risk of neonatal complications was increased only in offspring of women with schizophrenia. Crude ORs were also calculated for specific obstetric complications. Women with schizophrenia and bipolar disorder, but not women with unipolar depression, were significantly more likely to experience placental abnormalities, antepartum haemorrhages, and toxic side effects of drugs (alcohol, tobacco and illicit substances) compared with unaffected women. Children in all three diagnostic groups were more likely to experience fetal distress (significant in offspring of schizophrenia mothers), and to have a narcotic antagonist administered (significant in mothers with schizophrenia and unipolar depression). There was a non-significant tendency for a suboptimal 5-min Apgar score in the offspring of the women with schizophrenia, as well as for delayed respiration and intubation. There were no significant differences among the groups in mean gestational age, but newborns of the mothers with schizophrenia had a lower mean birthweight (3,248 g versus 3,334 g in the comparison group), a smaller head circumference (11.4% were in the lowest 10th percentile compared with 9.5% of the comparison offspring), and tended to be shorter (9.4% in the lowest 10th percentile compared with 8.7% of the offspring of unaffected mothers). A significantly larger proportion of babies of the mothers with schizophrenia (12.6% compared with 9.3% of the comparison group, 10.4% of the offspring of the mothers with unipolar disorder, and 9.9% of the offspring of the women with bipolar disorder) fell into the lowest 10th percentile of intrauterine growth using an index of percentage of expected birthweight based on whole-of-population birth data for Western Australia [16]. Smoking is known to have a marked impact on infant birthweight. When smoking in pregnancy was modelled, even assuming a worst-case scenario based on smoking data from the Australian national epidemiological survey of psychosis [17], it was not possible to explain all of the birthweight reduction in the offspring of mothers with schizophrenia. Finally, the risk of having any birth defect was similar for all children, regardless of their mother's psychiatric status. However, two categories of birth defects, cardiovascular defects (ventricular and atrial septal defects, patent ductus arteriosus, anomalies of the aorta) and minor physical anomalies, were significantly elevated in the children of women with schizophrenia.

Some 45% of the case children were born before their mother's first psychiatric admission. Multivariate analyses indicated that, relative to the comparison sample, mothers whose psychiatric illness commenced *before* the birth of their child were significantly more likely to experience obstetric complications (adjusted OR 1.1, 95% confidence interval, CI, 1.1–1.2). By contrast, mothers whose psychiatric illness commenced *after* the birth of their child were at no greater risk of obstetric complications than comparison mothers. However, there were a number of individual complications, specific to mothers with schizophrenia, where both pre- and post-onset rates were elevated. These included: abruption of the placenta; low birthweight; cardiovascular defects, and minor physical anomalies.

Summary

This study found that women with schizophrenia, bipolar disorder, and unipolar depression experience an increased overall incidence of obstetric complications, relative to the non-psychiatric comparison group. A major factor contributing to the increased risk of obstetric complications in women with schizophrenia and, to a lesser degree, in women with affective disorders appears to be the clustering of adverse maternal characteristics. Moreover, it is likely that behavioural disorganisation and environmental exposures including poor nutrition and substance use may play an important role in the excess of obstetric complications in births that occur after the onset of psychosis only. This is supported by the finding that the incidence of adverse outcomes in all three diagnostic groups was significantly increased only in pregnancies occurring after the onset of psychiatric illness. Some obstetric complications showed no pre-onset/post-onset differences. These occurred only in women with schizophrenia, and included placentation abnormalities, low birthweight, minor physical anomalies and cardiovascular birth defects. This constancy in rates suggests a diagnosis-specific pre-existing susceptibility that may involve both genetic and environmental components. The excess of cardiovascular birth defects, specific to mothers with schizophrenia, may involve parentally transmitted genes, some of which may be expressed in both the heart and brain or are in linkage disequilibrium with other genes conferring susceptibility to schizophrenia. It is also conjectured that low birthweight, occurring only in mothers with schizophrenia and with increased frequency in southern hemisphere springtime, may point to maternal exposure to winter infection resulting in a compromised intrauterine environment and restricted fetal growth.

Study 2. Adverse Pregnancy Outcome in Schizophrenia: Maternal and Paternal Influences

Prof. Christina Hultman and her team of researchers in the Department of Medical Epidemiology and Biostatistics at the Karolinska Institutet in Sweden are also working on pregnancy outcomes for women with schizophrenia. These women face an increased risk of adverse pregnancy outcomes: many smoke, misuse other substances and are socio-economically disadvantaged. Not only are these variables well-known risk factors for adverse pregnancy outcomes, but smoking is known to be causally related to fetal growth retardation and is also possibly associated with stillbirth and preterm birth. In several recent population-based studies, the increased risks for preterm birth, low birth weight and infant death in women with schizophrenia remained even after controlling for these covariates. Nonetheless, Prof. Hultman's research has previously shown that the increased risk for preterm birth among women with schizophrenia decreased from 70 to 40% when controlling for maternal factors [18]. Women who had had an episode of schizophrenia during their pregnancy had the highest risk of (a) preterm delivery (delivery before the 37th gestational week; OR 3.4, 95% CI 2.1–5.4), (b) low birth weight (birthweight under 2,500 g; OR 4.3, 95% CI 2.9–6.6), and (c) stillbirth (OR 4.4, 95% CI 1.4–13.8). Controlling for the high incidence of smoking during pregnancy among women with schizophrenia (51 vs. 24% in the normal population) and other maternal factors (single motherhood, maternal age, parity, maternal education, mothers' country of birth and pregnancy-induced hypertensive diseases) in a multiple regression model reduced the risk estimates markedly. However, even after adjustment, the risk for adverse pregnancy outcomes was generally doubled for women with an episode of schizophrenia during pregnancy compared with women in the control group. To test the hypothesis that infants of women with affective psychosis face increased risks of adverse pregnancy outcomes, Prof. Hultman and colleagues performed a similar study with data from the Swedish Medical Birth Register and the Hospital Discharge Register [19] on 5,593 births to mothers with affective psychoses and 46,068 control mothers. Births to mothers with affective psychosis had an increased risk of giving birth to preterm, small or growth-retarded babies. The risk for stillbirth and infant death during the 1st year of life was not significantly higher. The risks were greatest in mothers receiving hospital treatment for affective disorder during pregnancy (OR = 2.67, 95% CI 1.71–4.17). Affected mothers were approximately twice as likely to be heavy smokers than controls in this sample.

What might be the causes of the higher risk of preterm birth in women experiencing a psychotic episode during pregnancy? The risk may be partly accounted for by the use of antipsychotic medication during pregnancy, or by an increase in antipsychotic dosage during this period [20]. There has been some concern that antipsychotic medication may compromise uterine blood flow, and produce post-partum neonatal sedation and extrapyramidal signs. In a study following up the outcome of 215 pregnancies exposed to haloperidol or penfluridol in the first trimester, a higher rate of preterm births (13.9 vs. 6.9%, $p = 0.006$) was found compared with unexposed control pregnancies. The rate of congenital anomalies was not increased, but there were indications of fetal growth restriction, with a lower median birth weight for exposed full-term infants (3,250 vs. 3,415 g, $p = 0.004$). A second explanation may be that women with schizophrenia attend antenatal care clinics less frequently, and therefore may suffer from underdiagnosed medical conditions that affect their pregnancy outcomes. A third possibility is a common familial vulnerability for non-optimal pregnancy outcomes. Schizophrenia is under a high degree of genetic influence and offspring carry a 10% risk of developing schizophrenia themselves.

Less is known about the offspring's risk for adverse pregnancy outcomes if the father has schizophrenia. A recent study by Webb and colleagues found an increased risk for post-neonatal death among offspring of fathers with schizophrenia [21]. However, the risk was not statistically significant and a meta-analysis by Webb and colleagues indicated a need for more evidence on the effects of exposure to paternal disorder. Recent data based on a cohort of 1,890,550 births in Sweden indicate that, as well as offspring of mothers with schizophrenia, offspring of fathers with schizophrenia have an increased risk of non-optimal birth outcomes. It was shown that offspring of fathers with schizophrenia have increased risks for infant death, low birth weight, and being small for gestational age [22]. It is likely then that paternal schizophrenia also confers a risk for adverse pregnancy outcomes. Furthermore, since the risks were attenuated when covariates were included in the models, it is also possible to conclude that adverse pregnancy outcomes were not a consequence of schizophrenia in the parents, but rather a consequence of adverse socio-environmental conditions. However, the association between both maternal and paternal schizophrenia and infant death remained after controlling for these covariates. This could be attributed to environmental causes, but it is also possible that it is due to a common genetic effect.

In order to explore further the contribution of genetic factors to the associations reported, the Karolinska team studied the risk for adverse pregnancy outcomes and infant death among the full- and half-siblings of parents with schizophrenia. A genetic factor would be indicated if the risk was significantly higher for parental full siblings compared with parental half-siblings. Using the Swedish Multigeneration Register, we linked family information with the Swedish Hospital Discharge Register. For both siblings of mothers and siblings of fathers, there seems to be a difference in outcomes for full sisters compared with half-sisters. Thus, the researchers found a modest, but significantly increased risk of low birth weight if the parental full sibling had schizophrenia but not if the parental half-sibling had the disease, suggesting that, in addition to a maternal effect, there was a genetic mediation of the association between schizophrenia and low birthweight. On the other hand, the risk for infant death was not increased if the parent had a sibling with schizophrenia.

Summary

Not only do offspring of mothers with schizophrenia have an increased risk for non-optimal birth outcomes, but offspring of fathers with schizophrenia also are at increased risk for adverse outcomes. Much of the increased risk can be explained by maternal psychosocial factors such as smoking and single motherhood. In addition to non-optimal birth outcomes, the offspring of mothers and of fathers with schizophrenia also have double the risk of infant mortality. Excess mortality is largely attributable to post-neonatal death, but not sudden infant death syndrome, and cannot be explained by maternal behaviour during pregnancy. Consequently, it is important for researchers to focus on the mechanisms underlying these associations. This would include an examination of the effects of psychotropic medication on the developing fetus and the further work on the role of genetic risk factors. While a genetic risk factor may be implicated in the findings for birth weight, it does not appear to be indicated with respect to the findings for infant mortality.

Study 3. Linking Schizophrenia Epidemiology and Developmental Neurobiology: The Impact of Low Prenatal Vitamin D on Brain Development

It has been hypothesised by a research group led by Prof. John McGrath at The Queensland Centre for Mental Health Research and the Queensland Brain Institute, University of Queensland, that low prenatal vitamin D may be a risk factor for schizophrenia [23]. Many studies have shown that children born in winter and spring have a significantly increased risk of developing schizophrenia in later life [24]. Children born at higher latitudes are also at increased risk [25], with both the incidence and prevalence of schizophrenia being significantly greater in sites from higher latitudes [26]. Of interest, data from cold climates indicate that the incidence of schizophrenia is significantly higher in dark-skinned migrants compared with the native born [27]. Given that hypovitaminosis D is more common (a) during winter and spring, (b) at high latitudes, and (c) in dark-skinned individuals [28], low prenatal vitamin D 'fits' these key environmental features. Preliminary evidence from analytical epidemiology studies also suggests that prenatal vitamin D warrants closer attention as a candidate risk factor. For example, vitamin D supplements in the 1st year of life significantly reduced the risk of schizophrenia in males in a large Finnish Birth Cohort [29]. In addition, 25-hydroxyvitamin D_3 serum levels in 26 mothers whose children developed schizophrenia were numerically (but not significantly) lower than that of 51 control mothers whose children did not develop the disease [30]. There was a trend-level association between low maternal vitamin D levels and schizophrenia in a subgroup of children of African-American mothers [30], who would be at greatest risk of vitamin D deficiency because of their increased skin pigmentation. Access to larger samples of archival maternal and neonatal sera will be required to examine this hypothesis adequately. However, banked prenatal and/or neonatal samples of individuals who have developed schizophrenia are rare resources. Even if such samples were available, the ability to interpret associations from observational studies is an ever-present problem for epidemiology [31]. Randomized control trials of adverse prenatal exposures are clearly not an option. In the absence of alternative strategies, schizophrenia epidemiology is prone to become trapped in cycles of uninformative replications ('circular epidemiology') [32].

While analytical epidemiology is still a critical research tool, various types of 'translational epidemiology' can provide complementary research strategies. Linking exposures with genetic polymorphisms in genes that impact on relevant pathways is one such option [33]. Creating strong links between schizophrenia epidemiology and neuroscience is also a valuable research strategy. It can provide biological plausibility to candidate exposures, which can then feedback to refine measures of exposure or related research designs. Furthermore, the exploration of clues from epidemiology based on molecular, cellular and behavioural neuroscience can act as a catalyst for neuroscience discovery.

In the last few years, Prof. McGrath's group has been testing the hypothesis that prenatal vitamin D deficiency alters brain development and adult behaviour in rodent models. The experiments involve depleting female rats of vitamin D, mating them, and then repleting the dams with vitamin D immediately after birth. The offspring have been examined as neonates and as adults. The developmental vitamin D deficiency model (abbreviated as DVD model), is a developmental exposure only: from birth onwards, all maternal animals receive a diet containing normal levels of vitamin D. Full details of the dietary and breeding program are provided elsewhere [34].

The evidence shows that DVD deficiency leads to physiological alterations in pups at birth and into adulthood. DVD-deficient neonates had larger lateral ventricles, increased cell proliferation and reduced apoptosis, reduced density of neurotrophin receptor ($p75^{NTR}$) and reduced levels of nerve growth factor and glial cell line-derived growth factor compared with controls [35, 36]. As adults, these animals had larger lateral ventricles and reduced nerve growth factor expression compared with controls [37]. DVD-deficient adults had significant changes in protein and gene expression in the brain [38, 39]. From a behavioural perspective, DVD-deficient adults were more active than controls (that is, showing hyperlocomotion) [34, 40], and had altered attentional processing indicated by impaired latent inhibition [41]. Finally, some of the most robust and consistent findings in the DVD model have emerged in pharmacological studies. For example, the research group has shown that DVD-deficient rats have enhanced locomotion in response to the NMDA antagonist MK-801; a behaviour which is blocked by the neuroleptic haloperidol [34]. More recently, the DVD model has also been established in the mouse [42].

Summary

The animal experimental data have provided strong evidence demonstrating that DVD deficiency during gestation alters the trajectory of brain development in rodent models. Of particular interest are features of the DVD phenotype that are informative for schizophrenia research. Most notably, the increase in lateral ventricular volume is one of the most consistent neurobiological correlates of schizophrenia [43] and heightened sensitivity to dopaminergic and glutamatergic agents, which are neurotransmitters that are both thought to function abnormally in the brains of patients [44, 45]. Because hypovitaminosis D is prevalent in pregnant women [46], this research may have important public health implications. While the epidemiological evidence linking low prenatal vitamin D and schizophrenia remains inconclusive, the animal experiments have provided compelling evidence about the role of vitamin D in brain development. Thus, clues from schizophrenia epidemiology have uncovered previously unsuspected pathways involving vitamin D and brain development.

Conclusion

Exposures in utero and the early neonatal period may have short and long-term neuropsychiatric and other consequences. These risks may operate interactively with or independently of genetic risk for schizophrenia. This chapter outlined several approaches to the study of neurodevelopment and schizophrenia. In one approach, long-standing, prospective population-based health registers in Western Australia and Sweden have been employed as an efficient and cost-effective resource in the study of rare and complex diseases. Their use has helped advance our understanding of genetic and environmental factors associated with adverse obstetric and offspring outcomes in women with schizophrenia. Using a different approach to advance our understanding in this area of research, investigators in Queensland are working together with neuroscientists to evaluate the biological plausibility of candidate exposures [38], in this case vitamin D, using animal models.

Although genetic liability may account for some of the findings, all three research programs provide good evidence that maternal risk factors in pregnancy are also major determinants of adverse outcomes. Policies to ensure that mothers are well supported during their pregnancies so as to reduce risks related to poor outcomes are indicated. This is particularly the case for vulnerable women with severe mental illness. Data presented here go even further and suggest that paternal characteristics and psychopathology are also important in the antenatal period, and more attention needs to be paid to fathers during this critical time. The clinical significance of research into neurodevelopment and obstetric pathology in schizophrenia is not trivial. Women make up some 40% of the prevalent cases of psychosis, and about half the women will have children. More particularly, for the majority, schizophrenia onset is during their peak childbearing years [17]. Through this series of studies, the hope is that we will understand better both the consequences of sub-optimal uterine environments for the offspring of parents with schizophrenia, as well as the impact of neurodevelopmental disruption in utero on the risk of onset of schizophrenia in late adolescence-early adulthood for any child. The capacity to identify modifiable risk factors will not only improve outcomes for mothers and fathers with schizophrenia and their children, but also for the community at large.

References

1 Gottesman II, McGuffin P, Farmer AE: Clinical genetics as clues to the 'real' genetics of schizophrenia (a decade of modest gains while playing for time). Schizophr Bull 1987;13:23–44.

2 Gottesman II, Erlenmeyer-Kimling L: Family and twin strategies as a head start in defining prodromes and endophenotypes for hypothetical early-interventions in schizophrenia. Schizophr Res 2001;51:93–102.

3 Kraepelin E: Dementia Praecox and Paraphrenia (translated by R.M. Barclay). Edinburgh, Livingstone, 1919.

4 Weinberger DR: Implications of normal brain development for the pathogenesis of schizophrenia. Arch Gen Psychiatry 1987;44:660–669.

5 Marenco S, Weinberger D: The neurodevelopmental hypothesis of schizophrenia: following a trail of evidence from the cradle to the grave. Dev Psychopathol 2000;12:501–527.

6 Cannon M, Jones PB, Murray RM: Obstetric complications and schizophrenia: historical and meta-analytic review. Am J Psychiatry 2002;159:1080–1092.

7 Fish B: Neurobiologic antecedents of schizophrenia in children. Arch Gen Psychiatry 1977;34:1297–1313.

8 Weinberg SM, Jenkins EA, Marazita ML, Maher BS: Minor physical anomalies in schizophrenia: a meta-analysis. Schizophr Res 2007;89:72–85.

9 Ismail B, Cantor-Graae E, McNeil TF: Neurological abnormalities in schizophrenic patients and their siblings. Am J Psychiatry 1998;155:84–89.

10 Brown AS: Prenatal infection as a risk factor for schizophrenia. Schizophr Bull 2006;32:200–202.

11 Susser E, Neugebauer R, Hoek HW, Brown AS, Lin S, Labovitz D, et al: Schizophrenia after prenatal famine. Further evidence. Arch Gen Psychiatry 1996;53:25–31.

12 Barker DJP: Fetal and infant origins of adult disease. BMJ 1992;301:1111.

13 Jablensky A, Morgan V, Zubrick S, Bower C, Yellachich L: Pregnancy, delivery, and neonatal complications in a population cohort of women with schizophrenia and major affective disorders. Am J Psychiatry 2005;162:79–91.

14 McNeil T, Cantor-Graae E, Sjöström K: Obstetric complications as antecedents of schizophrenia: empirical effects of using different obstetric complication scales. J Psychiatr Res 1994;28:519–530.

15 McNeil TF, Sjöström K: McNeil-Sjöström Scale for Obstetric Complications. Version B: 95 03 01b. Malmö, Lund University, 1995.

16 Blair E: The undesirable consequences of controlling for birth weight in perinatal epidemiological studies. J Epidemiol Commun Health 1996;50:559–563.

17 Jablensky A, McGrath J, Herrman H, Castle D, Gureje O, Morgan V, et al: Psychotic disorders in urban areas: an overview of the methods and findings of the Study on Low Prevalence Disorders, National Survey of Mental Health and Wellbeing 1996–1998. Aust N Z J Psychiatry 2000;34:221–236.

18 Nilsson E, Lichtenstein P, Cnattingius S, Murray R, Hultman C: Women with schizophrenia: pregnancy outcome and infant death among their offspring. Schizophr Res 2002;58:221–229.

19 MacCabe JH, Martinsson L, Lichtenstein P, Nilsson E, Cnattingius S, Murray RM, et al: Adverse pregnancy outcomes in mothers with affective psychosis. Bipolar Disord 2007;9:305–309.

20 Pinkofsky HB: Psychosis during pregnancy: treatment considerations. Ann Clin Psychiatry 1997;9:175–179.

21 Webb R, Abel K, Pickles A, et al: Mortality in offspring of parents with psychotic disorders: A critical review and metal-analysis. Am J Psychiatry 2005; 162:1045–1056.

22 Nilsson E, Hultman CM, Cnattingius S, Otterblad Olausson P, Björk C, Lichtenstein P: Schizophrenia and offspring's risk for adverse pregnancy outcomes and infant death. Br J Psychiatry 2008;193:311–315.

23 McGrath J: Hypothesis: is low prenatal vitamin D a risk-modifying factor for schizophrenia? Schizophr Res 1999;40:173–177.

24 Torrey EF, Miller J, Rawlings R, Yolken RH: Seasonality of births in schizophrenia and bipolar disorder: a review of the literature. Schizophr Res 1997;28:1–38.

25 Saha S, Chant DC, Welham JL, McGrath JJ: The incidence and prevalence of schizophrenia varies with latitude. Acta Psychiatr Scand 2006;114:36–39.

26 Davies G, Welham J, Chant D, Torrey EF, McGrath J: A systematic review and meta-analysis of Northern Hemisphere season of birth studies in schizophrenia. Schizophr Bull 2003;29:587–593.

27 Cantor-Graae E, Selten JP: Schizophrenia and migration: a meta-analysis and review. Am J Psychiatry 2005;162:12–24.

28 Holick MF: Environmental factors that influence the cutaneous production of vitamin D. Am J Clin Nutr 1995;61:638S–645S.

29 McGrath J, Saari K, Hakko H, Jokelainen J, Jones P, Järvelin MR, Chant D, Isohanni M: Vitamin D supplementation during the first year of life and risk of schizophrenia: a Finnish birth cohort study. Schizophr Res 2004;67:237–245.

30 McGrath J, Eyles D, Mowry B, Yolken R, Buka S: Low maternal vitamin D as a risk factor for schizophrenia: a pilot study using banked sera. Schizophr Res 2003;63:73–78.

31 Davey Smith G, Ebrahim S: Epidemiology – is it time to call it a day? Int J Epidemiol 2001;30:1–11.

32 Kuller LH: Circular epidemiology. Am J Epidemiol 1999;150:897–903.

33 Smith GD, Ebrahim S: Mendelian randomization: prospects, potentials, and limitations. Int J Epidemiol 2004;33:30–42.

34 Kesby JP, Burne TH, McGrath JJ, Eyles DW: Developmental vitamin D deficiency alters MK 801-induced hyperlocomotion in the adult rat: an animal model of schizophrenia. Biol Psychiatry 2006;60:591–596.

35 Eyles D, Brown J, Mackay-Sim A, McGrath J, Feron F: Vitamin D3 and brain development. Neuroscience 2003;118:641–653.

36 Ko P, Burkert R, McGrath J, Eyles D: Maternal vitamin D(3) deprivation and the regulation of apoptosis and cell cycle during rat brain development. Brain Res Dev Brain Res 2004;153:61–68.

37 Feron F, Burne TH, Brown J, Smith E, McGrath JJ, Mackay-Sim A, et al: Developmental vitamin D3 deficiency alters the adult rat brain. Brain Res Bull 2005;65:141–148.

38 Almeras L, Eyles D, Benech P, Laffite D, Villard C, Patatian A, et al: Developmental vitamin D deficiency alters brain protein expression in the adult rat: implications for neuropsychiatric disorders. Proteomics 2007;7:769–780.

39 Eyles D, Almeras L, Benech P, Patatian A, Mackay-Sim A, McGrath J, et al: Developmental vitamin D deficiency alters the expression of genes encoding mitochondrial, cytoskeletal and synaptic proteins in the adult rat brain. J Steroid Biochem Mol Biol 2007; 103:538–545.

40 Burne TH, Becker A, Brown J, Eyles DW, Mackay-Sim A, McGrath JJ: Transient prenatal vitamin D deficiency is associated with hyperlocomotion in adult rats. Behav Brain Res 2004;154:549–555.

41 Becker A, Eyles DW, McGrath JJ, Grecksch G: Transient prenatal vitamin D deficiency is associated with subtle alterations in learning and memory functions in adult rats. Behav Brain Res 2005;161: 306–312.

42 Harms LR, Eyles DW, McGrath JJ, Mackay-Sim A, Burne TH: Developmental vitamin D deficiency alters adult behaviour in 129/SvJ and C57BL/6J mice. Behav Brain Res 2008;187:343–350.

43 Harrison PJ, Weinberger DR: Schizophrenia genes, gene expression, and neuropathology: on the matter of their convergence. Mol Psychiatry 2005;10:40–68.

44 Olney JW, Farber NB: Glutamate receptor dysfunction and schizophrenia. Arch Gen Psychiatry 1995;52:998–1007.

45 Laruelle M, Frankle WG, Narendran R, Kegeles LS, Abi-Dargham A: Mechanism of action of antipsychotic drugs: from dopamine D(2) receptor antagonism to glutamate NMDA facilitation. Clin Ther 2005;27(suppl A):S16–S24.

46 Hollis BW, Wagner CL: Vitamin D deficiency during pregnancy: an ongoing epidemic. Am J Clin Nutr 2006;84:273.

Dr. Vera A. Morgan
The University of Western Australia School of Psychiatry and Clinical Neurosciences
Level 3 Medical Research Foundation Building
Rear 50 Murray Street, Perth, WA 6000 (Australia)
Tel. +61 8 9224 0235/0290, Fax +61 8 9224 0285, E-Mail vmorgan@cyllene.uwa.edu.au

Newnham JP, Ross MG (eds): Early Life Origins of Human Health and Disease.
Basel, Karger, 2009, pp 205–217

Approaches to Evaluate Gene-Environment Interactions Underlying the Developmental Origins of Health and Disease

Craig E. Pennell[a,c] · Lyle J. Palmer[b] · Brian S. Knight[d,e] · Caroline Relton[f] · Stephen J. Lye[c,d,e]

[a]School of Women's and Infants' Health, and [b]Centre for Genetic Epidemiology and Biostatistics, The University of Western Australia, Perth, W.A., Australia; Departments of [c]Obstetrics and Gynaecology, and [d]Physiology, University of Toronto, [e]Samuel Lunenfeld Research Institute, Mount Sinai Hospital, Toronto, Ont., Canada; [f]Institute of Human Genetics, Newcastle University, Newcastle upon Tyne, UK

Research studies have established a clear relationship between antenatal and postnatal environments and the development of adult diseases including the metabolic syndrome (coronary heart disease, stroke, insulin resistance, type 2 diabetes and dyslipidemia), obesity, neurologic disorders and mental illness [1]. These observations have been confirmed in multiple human populations and in numerous animal studies in multiple species. It is clear that the environment of mother, baby and child is a key contributor to diseases and conditions that account for approximately one third of the global burden of disease in both developed and developing countries. Although adverse antenatal and postnatal environments increase the risk of particular adult diseases, not all individuals exposed to these environments develop these conditions, suggesting that an individual's genotype may contribute to the eventual outcome. Therefore, it has been suggested that gene-environment interactions underlie the developmental origins of health and disease (DOHaD).

Evidence of Gene-Environment Interactions Underlying Developmental Origins of Health and Disease

There is a growing body of evidence that both genetic and epigenetic pathways can mediate the relationship between the environment and the DOHaD (fig. 1). Variations in genetic sequence can influence the effect an adverse environment has on the risk of adult disease. Further, an adverse environment can induce epigenetic modification

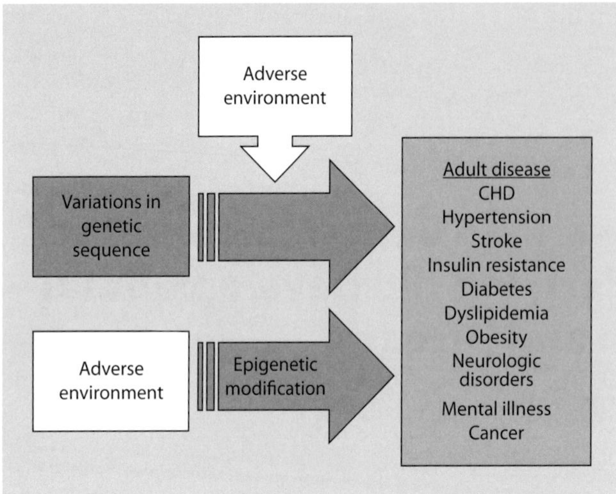

Fig. 1. Variations in genetic sequence can influence the effect an adverse environment has on the risk of adult disease. Further, an adverse environment can induce epigenetic modification of the genome which can influence an individual's risk of health and disease. Both of these mechanisms are likely to be responsible for the variation seen in an individual's response to an adverse environment. It is unlikely that these mechanisms are independent, and the relative contributions of each are yet to be clearly determined.

of the genome which can influence an individual's risk of health and disease. Both of these mechanisms are likely to be responsible for the variation seen in an individual's response to an adverse environment. It is unlikely that these mechanisms are independent, and the relative contributions of each are yet to be clearly determined.

Interaction between Variations in Genetic Sequence, the Environment and Developmental Origins of Health and Disease

There is a considerable body of literature attributing polymorphisms (particularly single nucleotide polymorphisms or SNPs) within genes as primary contributors of an individual's risk of disease. Much of the early work in DOHaD focused on relating population differences in disease risk and polymorphisms in key genes in the metabolic pathways regulating energy metabolism. Polymorphisms in the peroxisome proliferator-activated receptor (PPAR)-γ2 gene modify the relationship between size at birth and adult diseases such as insulin sensitivity and altered metabolism [2, 3], hypertension [4], obesity [5] and dyslipidemia [6]. The well-known association between small body size at birth and insulin resistance and/or hypertension was seen only in individuals with the high-risk Pro12Pro genotype and not the Pro12Ala variant in the PPARγ2. The Pro12Ala variant in PPARγ2 isoform-specific exon B is known to reduce the transcriptional activity of PPARγ2 [6]; therefore, PPARγ2 may just be a marker for DOHaD, and the polymorphism may not be directly inducing or controlling the expression of the phenotype. It may also be possible to see stronger associations with disease susceptibility if more than one interrelated marker can be identified.

Polymorphisms in the glucocorticoid receptor (GR), a key control element in the hypothalamic-pituitary-adrenal axis, have the potential to influence intrauterine glucocorticoid exposure during fetal growth and, perhaps, glucocorticoid metabolism and disease later in life. Indeed, polymorphisms in the GR have been implicated in determining obesity [7, 8], hypertension [9], hypercholesterolemia [9] and responses to psychosocial stress in adults [8]. Similarly, the haplotype structure of common variations within the GR gene locus has identified further genetic combinations linking birthweight to adult disease in the Helsinki Birth Cohort [10]. One of the haplotypes identified was associated with lower birth weight and length and higher fasting plasma and mean 24-hour salivary cortisol, indicating a potential association of body size at birth with later hypothalamic-pituitary-adrenal axis function. Furthermore, carriers of one of the haplotypes were found to have an increased association between body size at birth and impaired glucose tolerance or diabetes as adults in comparison with non-carriers [10]. Studies in non-human primates [11] show that polymorphisms in 5-HTT (serotonin transporter; 5-HTTLPR, which is also present in humans [12]) can alter behavioral vulnerability to modified early rearing environment. These observations suggest that complex interactions between genes and the early environment modulate developmental programming of adult disease.

Genetic differences not only regulate gene-environment interactions underlying disease onset but also regulate disease susceptibility following environmental alterations. The gut-brain regulatory peptide, ghrelin, is involved in controlling energy balance, and the Leu72Met polymorphism has been associated with impaired glucose tolerance and type 2 diabetes through defective first-phase insulin secretion [13]. In the Finnish Diabetes Prevention Study, where subjects with impaired glucose tolerance were randomized into either an intervention group (intensive diet and exercise intervention) or a control group, subjects with impaired glucose tolerance and the common Leu72Leu genotype developed type 2 diabetes less frequently under intervention circumstances ($OR = 0.28$, 95% CI 0.10–0.79; $p = 0.016$) than subjects with the Leu72Met allele [13]. Not only do polymorphic variations predispose individual to disease but they also impart a differential susceptibility to treatments in multifactorial diseases such as diabetes.

Interaction between the Environment, the Epigenome and Developmental Origins of Health and Disease

Epigenetics is an emerging scientific discipline which provides an attractive mechanism linking early life exposures with alterations in gene expression which may persist throughout life. There is growing evidence that both nutritional cues [14] and behavioral cues [15] during pregnancy and early postnatal life can influence the epigenomic state of a gene. Epigenetic changes can also be induced by lifestyle and environmental exposures including tobacco smoke, alcohol, chemical carcinogens, infectious agents and UV radiation [16]. Furthermore, some of these changes can persist across multiple generations [17].

Animal studies have shown that modulation of the methyl content of the maternal diet during pregnancy in mice results in altered DNA methylation patterns in offspring and altered gene expression [18]. Further evidence indicates that unbalanced prenatal nutrition induces persistent gene-specific (GR and PPARγ) changes in DNA methylation that alter mRNA expression in a rat model [19]. Moreover, investigators have recently provided evidence that in utero dietary manipulation affects gene expression and alters susceptibility to obesity in adulthood by altering the epigenome [20]. Furthermore, epigenetic regulation of 11β-hydroxysteroid dehydrogenase type 2 expression by DNA methylation has been demonstrated in vitro, providing a potential mechanism for programming of hypertension [21]. A number of human studies have also shown that manipulation of dietary folate causes detectable changes in global genomic DNA methylation status [22, 23], offering the possibility of long-term modifications in gene expression in response to diet.

Maternal behavioral cues have also been shown to influence the epigenomic state of a gene. Weaver et al. [15], utilizing a rat model, demonstrated that high levels of maternal caregiving behaviour (pup nursing, licking, and grooming) in the early postnatal period activate GR transcription and induce permanent GR hypomethylation. Manipulation of mother-offspring behavioral interactions in rats resulted in the alteration of the CpG methylation pattern of the I_7 GR promoter, and these postnatal changes in gene methylation subsequently lead to altered expression of the GR. This differential CpG methylation of specific GR promoter sites in the first exon (untranslated), which are spliced on to the common translated sequence beginning at exon 2, may also be implicated in the transmission of stable, long-term differences in GR expression over the life span of the rat [15]. Further, this maternal behaviour was shown to induce a permanent increase in hippocampal GR, although changes to the offspring hippocampal transcriptome could subsequently be reversed in adulthood by histone deacetylase administration [24].

Emerging evidence suggests that nutritional manipulations can epigenetically mark the genome and impact gene expression over the life course. The dietary factors involved, the genes that are susceptible and the developmental plasticity of the effects remain to be determined. Variable penetrance is a recognised feature of disease-associated SNPs, and it is plausible that epigenetic regulation of gene expression may explain a proportion of such variable penetrance. Thus, genetic variation and epigenetic variation may act in concert to modulate disease risk. Integrated approaches, such as the common disease genetic and epigenetic model [25] will be useful tools in exploring this further.

Approaches and Challenges for Investigating Gene-Environment Interactions and Developmental Origins of Health and Disease

In the post-genomic era, there are many approaches that one can take to investigate gene-environment interactions underlying DOHaD. Broadly, these can be grouped into: (1) animal approaches, which offer the advantage of well-defined, often inbred strains, with relatively short life spans, and (2) human studies, which require large, well-phenotyped cohorts which can take decades to establish. The strengths and challenges of each of these approaches will be discussed in detail.

Animal Approaches

The laboratory mouse has been used extensively to analyse complex genetic traits but only recently have mice been used in the study of DOHaD. Complex genetic traits can be scrutinized in well-defined inbred mouse strains because genetic factors have been segregated and fixed during inbreeding and environmental effects can be controlled [26]. Comparison of mouse strains using chromosomal maps of SNPs can reveal inbred or recombinant strains that will be phylogenetically distant due to separation by major bifurcations in phylogenetic hierarchical trees. This type of analysis has revealed that three of the more common strains (C57BL/6J, A/J or C3H/HeJ) can be genetically separated into distinct groups [27]. Of particular importance for DOHaD research, these three genetically distinct mouse strains differ in their physiologic characteristics and their responses to changes in their environment.

The mouse strains A/J and C57BL/6J differ in many important physiological characteristics including homocysteine [28] and anxiety levels [29]. Additionally, exposure of C57BL/6J to a high-fat diet during adolescence and adulthood induces features of the metabolic syndrome (obesity and diabetes); however, the same high-fat diet had little effect on the A/J strain [26]. Similarly, strain differences can also be seen with the comparison of the phylogenetically distant C3H/HeJ and C57BL/6J which differ in HDL metabolism and atherogenesis when challenged with a high-fat, high-cholesterol diet [30]. Thus, comparison of complex traits in genetically distinct mice, using models of DOHaD, could provide novel insights into the gene-environment interactions that underlie DOHaD. Results from genetic screens of mouse models may indicate target chromosomal regions and/or candidate genes that can subsequently be tested in population or family studies in humans.

Recently, our research team has demonstrated remarkable strain differences between A/J and C57BL/6J in response to an adverse intrauterine environment induced by maternal dietary restriction (DR) during pregnancy [31, 32]. The maternal DR entailed a strain-specific, 30% reduction in total calorie intake between day 6.5 and day 18.5 of gestation, which induced a 15% reduction in birthweight of the pups [32]. Maternal DR had the greatest impact on C57BL/6J mice compared with A/J as demonstrated by significant alterations in fetal organ weights (liver, kidneys and placenta) and significant increases in maternal glucocorticoids by the end of pregnancy [32]. Further, in postnatal life C57BL/6J offspring demonstrated postnatal catch-up growth, increased total body weight, decreased lean mass, increased body fat and increased glucose intolerance, compared with the A/J offspring at 12 and 26 weeks of age [31]. Thus, from these studies it appears that the C57BL/6J responds in similar ways to rats [33], guinea pigs [34] and humans [35], where the impact of adverse environments has been evaluated. In contrast, the A/J strain appeared relatively resistant to the impact of these environmental challenges. Therefore inter-strain comparisons between C57BL/6J and A/J offer a unique opportunity to gain a better understanding of the genetic (and epigenetic) basis for gene-environment interactions underlying DOHaD.

The differing responses of the well-defined inbred mouse strains C57BL/6J and A/J to antenatal undernutrition and postnatal ad libitum or high-fat diets have afforded our research team the opportunity to investigate the relative contributions of (1) genetic and epigenetic factors and (2) environmental factors to the subsequent development of specific phenotypes including obesity, impaired glucose tolerance, hypertension and cardiac function [unpubl.]. Contributions of genetic and environmental factors to outcomes were assessed via generalized linear modelling with repeated measures and the contribution of genetics and the environment was summarised as the proportion of variance explained in the model (partial r^2). These data (table 1) demonstrate a wide variation in the contribution of genetic/epigenetic factors (2–43%) and the environment (2–18%) to unfavorable outcomes in response to adverse environments during the antenatal and postnatal periods which are dependent on the specific phenotype assessed. Moreover, these data provide strong support for a significant contribution of genetic/epigenetic factors to the risk of developing obesity, impaired glucose tolerance and hypertension.

A number of different approaches are available to utilize well-defined inbred mouse strains to determine the genetic basis for specific complex phenotypes. Segregation of quantitative trait loci (QTL) is evident in the substantial genetic variation seen in inbred mouse strains. Traditional approaches to QTL identification can be tedious and time consuming due to substantial backcrossing required; however, two novel methods have expedited this process. The first involves prior construction of a panel of chromosome substitution strains (CSS), where the host strain carries all but one of its own chromosomes and the chromosome of interest is supplied by a donor strain [26, 36]. Since the replacement of each chromosome on the host background in CSS creates 22 sub-strains, only minimal backcrossing is required to localize QTLs. Using this approach, 150 QTL have already been associated with alterations in serum levels of sterols and amino acids, diet-induced obesity, and anxiety in postnatal overnutrition studies [26, 36]. Alternatively, Emerillon Therapeutics Inc. has created a panel of 37 independent recombinant congenic strains (RCS) using two strains as both host and donor. These RCS strains have been genotyped for a total of 625 informative

Table 1. The contribution of genetics and the environment summarized as the proportion of variance explained in the model (partial r^2)

Outcome[1]	Base[ζ]	Base[ζ] + strain	Base[ζ] + strain + AN nutrition	Base[ζ] + strain + PN nutrition	Base[ζ] + strain + AN + PN nutrition
Body weight	19	25 (+6)	31 (+12)	36 (+17)	43 (+24)
Percent body fat	3	46 (+43)	48 (+45)	52 (+49)	55 (+52)
Area under curve glucose tolerance test	9	28 (+19)	29 (+20)	29 (+20)	30 (+21)
Area under curve insulin	2	10 (+8)	14 (+12)	10 (+8)	14 (+12)
Systolic blood pressure	2	25 (+23)	28 (+26)	27 (+25)	31 (+28)
Tei Index	33	35 (+2)	38 (+5)	36 (+3)	39 (+6)

Figures indicate percentages. AN = Antenatal; PN = postnatal. Base[ζ] controls for mother, number in litter and sex of pup.

[1] Body weight, body fat determined by DEXA scanning and glucose tolerance testing were performed at 26 weeks of age, whilst blood pressure and cardiac assessments were performed at 12 weeks of age.

microsatellite DNA markers covering the entire genome, with an average spacing of 2.6 cM and display excellent coverage of congenic segments from both parental genomes. Previously, these RCS mice have provided further understanding into the genetic complexity controlling susceptibility to diseases, such as hyperlipidemia [36], lung cancer [37] and infection [38]. Both the CSS and RCS mice can be used to map simple and complex traits in a genome-wide fashion and thus will allow researchers to systematically probe many of the gene-environment interactions believed to contribute to DOHaD pathogenesis.

Much like QTLs, there is substantial genetic variation in SNPs that can be assessed in inbred mouse strains. Evaluation of gene polymorphisms in key metabolic pathways regulating energy metabolism supports a role for gene-environment interactions in DOHaD. Polymorphisms in genes such as the PPARγ2 and GR have been implicated in susceptibility modification to adult diseases including insulin sensitivity and metabolism [2, 3], hypertension [4, 9], obesity [5, 7, 8], hypercholesterolemia [9] and dyslipidemia [6]. Comparison of polymorphisms within mouse strains could potentially identify specific genes with central roles in DOHaD; however, translating knowledge of specific genes to clinical outcomes is not necessarily straightforward, since most SNPs do not alter function [39].

Singular (forward or reverse) genetic approaches to identify genes involved in complex trait diseases have had limited success. Recent research [27, 40] has developed an integrative genetic approach that appears to be considerably more successful. This hybrid procedure is heavily dependent on bioinformatics and in silico analysis. Using these integrative techniques, 5-lipoxygenase was recently identified as a susceptibility gene for obesity and bone traits. This finding was achieved by integrating information of differential expression of liver genes (determined by microarray) with high frequency SNPs to clinically relevant QTL [40]. Further evidence was provided by the use of gene ontology of biological process categories associated with obesity-related traits and expression mapping after treatment with pharmaceutical agonists that interfered with liver function. In a similar manner, *Insig2* was identified as a candidate susceptibility gene for plasma cholesterol levels by first creating a high-density SNP map between murine strains and then integrating in silico QTL mapping with progressive QTL mapping strategies, including Bayesian network modeling to determine interacting gene pathways, in a segregating mouse population [27]. These examples demonstrate the potential of coupling murine models to integrated genetic analysis to enable the discovery of specific gene-environment interactions more rapidly than has ever been possible before. These are indeed exciting developments with far-reaching implications for DOHaD research.

Animal studies also offer the potential to interrogate the role of epigenetic mechanisms in the DOHaD. Analysis of multiple tissue types with specific epigenetic signatures at multiple time points across the life course from fetal life to adulthood provide unprecedented opportunities to analyse epigenetic patterns that are simply not possible in human studies. Rodent models have been used in a number of studies to date to explore epigenetic factors in nutritional programming of disease in later life [14, 19, 20]. Further work in this area will include the application of microarray-based technologies to discover novel targets showing differential epigenetic patterns induced by early life exposures [41].

Human Approaches

There are a number of important issues to consider when utilizing human cohorts to study genetic associations with the DOHaD. The first essential requirement to conduct genetic association studies to decipher a complex disease is a clear definition of the outcome: the phenotype [41]. Further, detailed environmental data are required to attempt to unravel the complex interactions between genes and the environment. Third, the issue of sample size requires careful consideration [42]. To choose the appropriate sample size and genotyping platform, power calculations are required that take into account the genetic model, tag SNP selection, and the population of interest [43]. Most genome-wide association studies (GWAS) now utilize a minimum of several thousand cases and controls to evaluate complex diseases such as diabetes [44], hypertension [45] and coronary artery disease [45]. Further, meta-analyses and pooling of GWAS data are producing sample sizes of the order of 100,000 cases for some diseases. Sample sizes of this magnitude are relatively straightforward to acquire when analysing cross-sectional data; however, there are few cohorts that have well-characterised, longitudinal phenotypic and environmental data acquired from birth and followed up over many decades that will be required to study the gene-environment interactions underlying DOHaD.

Another issue that is of particular relevance to the study of gene-environment interactions underlying DOHaD is whose DNA is important: the child, mother, father or a combination of all three members of the family triad? Currently our ability to genotype is more advanced than the complex statistical tools required to analyze the relationship of two (or three genotypes) to complex phenotypic outcomes. To further complicate data analyses, it is likely that complex diseases result from the interplay of multiple genes and the environment. These multi-locus interactions further complicate power calculations and statistical analyses [46]. Epistasis, when one or more genes mask or suppress the effects of another gene (or genes) [47], can result in failure to detect the effects of a single locus if one only considers one gene and no effects. Further, epistasis can result in the detection of a single locus effect in one study but not in others [47].

Over the last 5 years, there has been increasing utilization of human cohorts to study gene-environment interactions underlying DOHaD utilizing a candidate gene approach (see previous section for detail). In brief, polymorphisms in the PPARγ2 gene have been shown to modify the relationship between size at birth and adult diseases such as insulin sensitivity and altered metabolism [2, 3], hypertension [4], obesity [5] and dyslipidemia [6]. Similarly, polymorphisms in the GR have been implicated in determining obesity [7, 8, 48], hypertension [9], hypercholesterolemia [9] and responses to psychosocial stress in adults [8]. Moreover, human cohorts have also been utilized to demonstrate that genetic differences regulate disease susceptibility following environmental alterations, with polymorphisms in ghrelin being shown to influence the rate of progression from impaired glucose tolerance to type 2 diabetes (in response to an intensive diet and exercise intervention) in the Finnish Diabetes Prevention Study [13].

Candidate gene approaches are rapidly being replaced by GWAS to unravel the genetic basis of many complex diseases including: type 1 [45] and type 2 diabetes [44, 45], Crohn's disease [45], bipolar disorder [45], coronary artery disease [45], rheumatoid arthritis [45], hypertension [45], multiple sclerosis [49], macular degeneration [50], autism [51] and schizophrenia [52]. Three key advances have created unprecedented opportunities for understanding the pathogenic basis of common human diseases: (1) extensive catalogues of DNA sequence variants across the human genome have been compiled; (2) dramatic progress in sequencing and SNP genotyping technologies, and (3) increasing availability of large-scale, population-based human samples [53]. Both linkage and candidate gene approaches have generally failed to discover replicable genetic associations with chronic disease phenotypes [47, 54]. In contrast, GWAS provide an unbiased approach to identify genomic regions (coding and non-coding sequences) containing genetic variants causally associated with disease phenotypes [45]. The current focus on GWAS has stemmed partly from the realization that candidate-gene association studies of complex phenotypes often either fail to discover susceptibility loci or they produce results which are not reproduced in replication studies [47, 54]. By enabling hypothesis-free research, GWAS have evolved a powerful new paradigm to understand complex disease biology [45]. It is anticipated that GWAS will offer many new insights into the gene-environment interactions underlying DOHaD, with results of the first GWAS addressing DOHaD expected in 2009.

Human epigenetic epidemiology is a field in its infancy. There is little in the way of population level data cataloguing epigenetic variation; consequently, the role of epigenetic variation in common complex diseases can only be speculated. As mentioned above, an integrated approach which considers both SNPs and epigenetic variation will be important in increasing our understanding of gene-environment interactions. Knowledge of environmental exposures, particularly in early life, that persistently alter the epigenotype will inevitably develop as research momentum in this area builds.

Conclusions

We now recognize that environmental influences alone do not determine an individual's likelihood of developing health and disease; rather, interactions between an individual's genetic makeup and the environment underlie DOHaD. While these are exciting times for DOHaD research, there remain significant clinical, scientific and ethical issues that need to be addressed before we can expect to reduce the incidence of DOHaD outcomes such as the metabolic syndrome. Clinical phenotyping needs to be much more precise if we are to make most effective use of the wealth of genomic information. Progress will require detailed clinical dissection of disease and/or the identification of endophenotypes. The vast amounts of genetic and epidemiologic data that are being generated by large-scale studies have created a remarkable challenge to our analytic tools and statistical models. Ethical issues regarding the use of these data also need to be addressed. Nevertheless, there is compelling evidence that diseases and conditions that represent some of today's most important global public health issues result from interactions between environmental influences and an individual's genetic makeup. Identification of the mechanistic basis for these interactions will have a tremendous impact on our ability to develop lifestyle or medical intervention strategies aimed at preventing the adverse outcomes associated with developmental origins of health and adult disease. The potential to impact on global disease burden and the opportunity to establish healthy life-long trajectories for women and their infants are immense.

References

1 McMillen IC, Robinson JS: Developmental origins of the metabolic syndrome: prediction, plasticity, and programming. Physiol Rev 2005;85:571–633.

2 Eriksson JG, Lindi V, Uusitupa M, et al: The effects of the Pro12Ala polymorphism of the peroxisome proliferator-activated receptor-gamma2 gene on insulin sensitivity and insulin metabolism interact with size at birth. Diabetes 2002;51:2321–2324.

3 Laakso M: Gene variants, insulin resistance, and dyslipidaemia. Curr Opin Lipidol 2004;15:115–120.

4 Yliharsila H, Eriksson JG, Forsen T, et al: Interactions between peroxisome proliferator-activated receptor-gamma 2 gene polymorphisms and size at birth on blood pressure and the use of antihypertensive medication. J Hypertens 2004;22:1283–1287.

5 Pihlajamaki J, Vanhala M, Vanhala P, Laakso M: The Pro12Ala polymorphism of the PPAR gamma 2 gene regulates weight from birth to adulthood. Obes Res 2004;12:187–190.

6 Eriksson JG, Forsen TJ, Osmond C, Barker DJ: Pathways of infant and childhood growth that lead to type 2 diabetes. Diabetes Care 2003;26:3006–3010.

7 Rosmond R: Association studies of genetic polymorphisms in central obesity: a critical review. Int J Obes Relat Metab Disord 2003;27:1141–1151.

8 Wust S, Van Rossum EF, Federenko IS, Koper JW, Kumsta R, Hellhammer DH: Common polymorphisms in the glucocorticoid receptor gene are associated with adrenocortical responses to psychosocial stress. J Clin Endocrinol Metab 2004;89:565–573.

9 Di Blasio AM, van Rossum EF, Maestrini S, et al: The relation between two polymorphisms in the glucocorticoid receptor gene and body mass index, blood pressure and cholesterol in obese patients. Clin Endocrinol (Oxf) 2003;59:68–74.

10 Rautanen A, Eriksson JG, Kere J, et al: Associations of body size at birth with late-life cortisol concentrations and glucose tolerance are modified by haplotypes of the glucocorticoid receptor gene. J Clin Endocrinol Metab 2006;91:4544–4551.

11 Champoux M, Bennett A, Shannon C, Higley JD, Lesch KP, Suomi SJ: Serotonin transporter gene polymorphism, differential early rearing, and behavior in rhesus monkey neonates. Mol Psychiatry 2002;7:1058–1063.

12 Arbelle S, Benjamin J, Golin M, Kremer I, Belmaker RH, Ebstein RP: Relation of shyness in grade school children to the genotype for the long form of the serotonin transporter promoter region polymorphism. Am J Psychiatry 2003;160:671–676.

13 Mager U, Lindi V, Lindstrom J, et al: Association of the Leu72Met polymorphism of the ghrelin gene with the risk of Type 2 diabetes in subjects with impaired glucose tolerance in the Finnish Diabetes Prevention Study. Diabet Med 2006;23:685–689.

14 Waterland RA, Dolinoy DC, Lin JR, Smith CA, Shi X, Tahiliani KG: Maternal methyl supplements increase offspring DNA methylation at Axin Fused. Genesis 2006;44:401–406.

15 Weaver IC, Cervoni N, Champagne FA, et al: Epigenetic programming by maternal behavior. Nat Neurosci 2004;7:847–854.

16 Herceg Z: Epigenetics and cancer: towards an evaluation of the impact of environmental and dietary factors. Mutagenesis 2007;22:91–103.

17 Bertram C, Khan O, Ohri S, Phillips DI, Matthews SG, Hanson MA: Transgenerational effects of prenatal nutrient restriction on cardiovascular and hypothalamic-pituitary-adrenal function. J Physiol 2008;586:2217–2229.

18 Waterland RA: Do maternal methyl supplements in mice affect DNA methylation of offspring? J Nutr 2003;133:238; author reply 239.

19 Lillycrop KA, Phillips ES, Jackson AA, Hanson MA, Burdge GC: Dietary protein restriction of pregnant rats induces and folic acid supplementation prevents epigenetic modification of hepatic gene expression in the offspring. J Nutr 2005;135:1382–1386.

20 Dolinoy DC, Weidman JR, Waterland RA, Jirtle RL: Maternal genistein alters coat color and protects Avy mouse offspring from obesity by modifying the fetal epigenome. Environ Health Perspect 2006;114:567–572.

21 Alikhani-Koopaei R, Fouladkou F, Frey FJ, Frey BM: Epigenetic regulation of 11 beta-hydroxysteroid dehydrogenase type 2 expression. J Clin Invest 2004;114:1146–1157.

22 Friso S, Choi SW, Girelli D, et al: A common mutation in the 5,10-methylenetetrahydrofolate reductase gene affects genomic DNA methylation through an interaction with folate status. Proc Natl Acad Sci U S A 2002;99:5606–5611.

23 Mathers JC: Reversal of DNA hypomethylation by folic acid supplements: possible role in colorectal cancer prevention. Gut 2005;54:579–581.

24 Weaver IC, Meaney MJ, Szyf M: Maternal care effects on the hippocampal transcriptome and anxiety-mediated behaviors in the offspring that are reversible in adulthood. Proc Natl Acad Sci U S A 2006;103:3480–3485.

25 Bjornsson HT, Fallin MD, Feinberg AP: An integrated epigenetic and genetic approach to common human disease. Trends Genet 2004;20:350–358.
26 Singer JB, Hill AE, Burrage LC, et al: Genetic dissection of complex traits with chromosome substitution strains of mice. Science 2004;304:445–448.
27 Cervino AC, Li G, Edwards S, et al: Integrating QTL and high-density SNP analyses in mice to identify Insig2 as a susceptibility gene for plasma cholesterol levels. Genomics 2005;86:505–517.
28 Ernest S, Hosack A, O'Brien WE, Rosenblatt DS, Nadeau JH: Homocysteine levels in A/J and C57BL/6J mice: genetic, diet, gender, and parental effects. Physiol Genomics 2005;21:404–410.
29 Singer JB, Hill AE, Nadeau JH, Lander ES: Mapping quantitative trait loci for anxiety in chromosome substitution strains of mice. Genetics 2005;169:855–862.
30 Machleder D, Ivandic B, Welch C, Castellani L, Reue K, Lusis AJ: Complex genetic control of HDL levels in mice in response to an atherogenic diet. Coordinate regulation of HDL levels and bile acid metabolism. J Clin Invest 1997;99:1406–1419.
31 Knight BS, Pennell CE, Adamson SL, Lye SJ: The impact of murine strain and gender on postnatal development after maternal dietary restriction during pregnancy. J Physiol 2007;581:873–881.
32 Knight BS, Pennell CE, Shah R, Lye SJ: Strain differences in the impact of dietary restriction on fetal growth and pregnancy in mice. Reprod Sci 2007;14:81–90.
33 Desai M, Gayle D, Babu J, Ross MG: Programmed obesity in intrauterine growth-restricted newborns: modulation by newborn nutrition. Am J Physiol Regul Integr Comp Physiol 2005;288:R91–R96.
34 Kapoor A, Dunn E, Kostaki A, Andrews MH, Matthews SG: Fetal programming of hypothalamo-pituitary-adrenal function: prenatal stress and glucocorticoids. J Physiol 2006;572:31–44.
35 Flanagan DE, Moore VM, Godsland IF, Cockington RA, Robinson JS, Phillips DI: Fetal growth and the physiological control of glucose tolerance in adults: a minimal model analysis. Am J Physiol Endocrinol Metab 2000;278:E700–E706.
36 Castellani LW, Weinreb A, Bodnar J, et al: Mapping a gene for combined hyperlipidaemia in a mutant mouse strain. Nat Genet 1998;18:374–377.
37 Fijneman RJ, de Vries SS, Jansen RC, Demant P: Complex interactions of new quantitative trait loci, Sluc1, Sluc2, Sluc3, and Sluc4, that influence the susceptibility to lung cancer in the mouse. Nat Genet 1996;14:465–467.
38 Tuite A, Elias M, Picard S, Mullick A, Gros P: Genetic control of suceptibility to Candida albicans in susceptible A/J and resistant C57BL/6J mice. Genes Immun 2005;6:672–682.
39 Chakravarti A: It's raining SNPs, hallelujah? Nat Genet 1998;19:216–217.
40 Mehrabian M, Allayee H, Stockton J, et al: Integrating genotypic and expression data in a segregating mouse population to identify 5-lipoxygenase as a susceptibility gene for obesity and bone traits. Nat Genet 2005;37:1224–1233.
41 Esteller M: Cancer epigenomics: DNA methylomes and histone-modification maps. Nat Rev Genet 2007;8:286–298.
42 Ziegler A, Konig IR, Thompson JR: Biostatistical aspects of genome-wide association studies. Biomed J 2008;50:8–28.
43 Klein RJ: Power analysis for genome-wide association studies. BMC Genet 2007;8:58.
44 Horikawa Y, Miyake K, Yasuda K, et al: Replication of genome-wide association studies of type 2 diabetes susceptibility in Japan. J Clin Endocrinol Metab 2008;93:3136–3141.
45 Genome-wide association study of 14,000 cases of seven common diseases and 3,000 shared controls. Nature 2007;447:661–678.
46 Pickrell J, Clerget-Darpoux F, Bourgain C: Power of genome-wide association studies in the presence of interacting loci. Genet Epidemiol 2007;31:748–762.
47 Williams SM, Haines JL, Moore JH: The use of animal models in the study of complex disease: all else is never equal or why do so many human studies fail to replicate animal findings? Bioessays 2004;26:170–179.
48 van Rossum EF, Voorhoeve PG, te Velde SJ, et al: The ER22/23EK polymorphism in the glucocorticoid receptor gene is associated with a beneficial body composition and muscle strength in young adults. J Clin Endocrinol Metab 2004;89:4004–4009.
49 Oksenberg JR, Baranzini SE, Sawcer S, Hauser SL: The genetics of multiple sclerosis: SNPs to pathways to pathogenesis. Nat Rev Genet 2008;9:516–526.
50 Edwards AO, Lee SJ, Fridley BL, Tosakulwong N: Density of common complex ocular traits in the aging eye: analysis of secondary traits in genome-wide association studies. PLoS ONE 2008;3:e2510.
51 Liu XQ, Paterson AD, Szatmari P: Genome-wide linkage analyses of quantitative and categorical autism subphenotypes. Biol Psychiatry 2008;64:561–570.

52 Fanous AH, Neale MC, Webb BT, et al: A genome-wide scan for modifier loci in schizophrenia. Am J Med Genet B Neuropsychiatr Genet 2007;144B:589–595.
53 Davey Smith G, Ebrahim S, Lewis S, Hansell AL, Palmer LJ, Burton PR: Genetic epidemiology and public health: hope, hype, and future prospects. Lancet 2005;366:1484–1498.
54 Chanock SJ, Manolio T, Boehnke M, et al: Replicating genotype-phenotype associations. Nature 2007;447:655–660.

Dr. Craig E. Pennell
School of Women's and Infants' Health, M550
The University of Western Australia, 2nd Floor, A Block, King Edward Memorial Hospital
Perth, WA 6008 (Australia)
Tel. +61 8 9340 1326, Fax +61 8 9381 3031, E-Mail craig.pennell@uwa.edu.au

Author Index

Subject Index